IN VIVO THE CULTURAL MEDIATIONS OF BIOMEDICAL SCIENCE

Phillip Thurtle and Robert Mitchell, Series Editors

IN VIVO THE CULTURAL MEDIATIONS OF BIOMEDICAL SCIENCE

In Vivo: The Cultural Mediations of Biomedical Science is dedicated to the interdisciplinary study of the medical and life sciences, with a focus on the scientific and cultural practices used to process data, model knowledge, and communicate about biomedical science. Through historical, artistic, media, social, and literary analysis, books in the series seek to understand and explain the key conceptual issues that animate and inform biomedical developments.

The Transparent Body
A Cultural Analysis of Medical Imaging
by José van Dijck

José van Dijck

THE TRANSPARENT BODY

A CULTURAL ANALYSIS OF MEDICAL IMAGING

UNIVERSITY OF WASHINGTON PRESS • SEATTLE AND LONDON

This book is published with the assistance of a grant from the McLellan Endowed Series Fund, established through the generosity of Martha McCleary McLellan and Mary McLellan Williams.

Printed in Canada
Design by Echelon Design
10 09 08 07 06 05 5 4 3 2 1

University of Washington Press
P.O. Box 50096, Seattle, WA 98145, U.S.A.
www.washington.edu/uwpress

Library of Congress Cataloging-in-Publication Data

Dijck, José van.
The transparent body : a cultural analysis of medical imaging / by José van Dijck.
p. ; cm.—(In vivo)
Includes bibliographical references and index.
ISBN 0-295-98490-2 (pbk. : alk. paper)
1. Diagnostic imaging—Social aspects. 2. Diagnostic imaging—History.
3. Medicine and the humanities.
[DNLM: 1. Diagnostic Imaging—history. 2. Human Body. 3. Mass Media—trends. WN 11.1 D575T 2004] I. Title. II. In vivo (Seattle, Wash.)
RC78.7.D53D553 2004
616.07'54—dc22 2004023548

The paper used in this publication is acid-free and 90 percent recycled from at least 50 percent post-consumer waste. It meets the minimum requirements of American National Standard for Information Sciences-Permanence of Paper for Printed Library Materials, ANSI Z39.48-1984.

Cover photograph by Lautaro Gabriel Gonda

TO MY MOTHER

CONTENTS

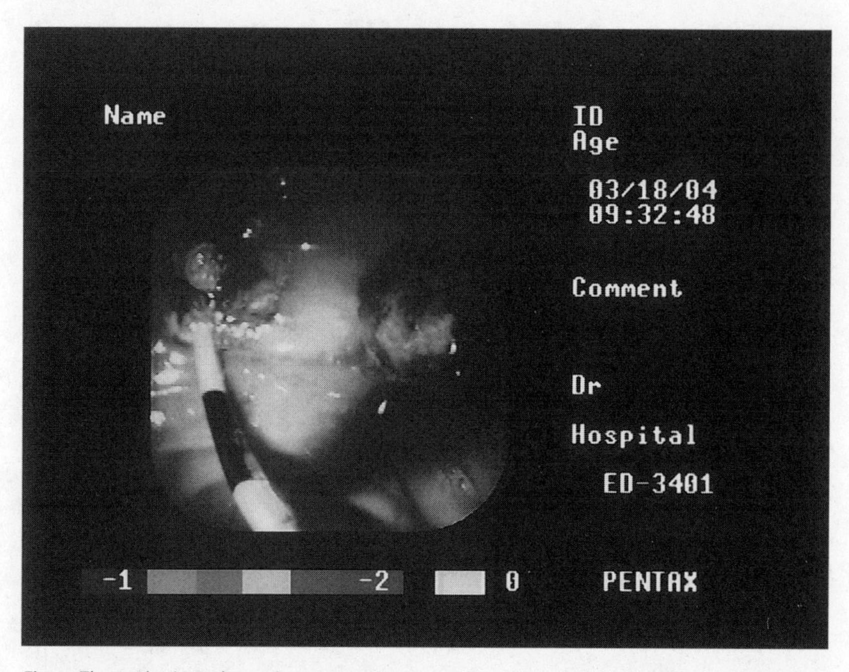

Fig. 1. The author's endoscopic surgery. Courtesy of Dr. Koek and Dr. Kamphuis, Academic Hospital, Maastricht.

PREFACE

The growing importance of medical imaging for our personal and collective experience of illness and pain may be illustrated by an experience I had while writing this book. For several weeks I had been experiencing sharp pain attacks in my upper stomach, accompanied by fits of nausea. When my general practitioner finally referred me to the hospital for a gastroscopy and an ultrasound, I had gradually come to believe that my symptoms signaled a bilious disorder. However, since my gall bladder had been removed twenty-three years ago, a recurrence of stones is highly unlikely, so my physician did not subscribe to my self-diagnosis. Neither the gastroscopy nor the ultrasound turned up anything to confirm my suspicion. Listening to my desperation, the gastroenterologist ordered a blood test, just to make sure. Upon my return from the hospital, I began to doubt my own symptoms; I decided my worries were groundless and my GP was right: the images showed nothing, so I should go back to work.

The next day, a phone call completely changed my perspective: the gastroenterologist summoned me to check in to the hospital immediately, without delay, because my blood tests for the liver and pancreas signaled a serious obstruction. Since I was at high risk for pancreatitis (a potentially life-threatening infection of the pancreas), I had to undergo an emergency ERCP—an endoscopic operation of the gall and pancreas ducts. This is a very high-tech procedure, in which a camera is inserted into the gall ducts through a tube in the esophagus; after injecting a contrast liquid into the ducts and making an X ray, doctors can localize obstacles in even the tiniest of tracts. Through that same tube, they may subsequently insert an instrument to remove obstructions. To my delight, the operating specialists captured two green stones that were responsible for all my pain and suffering! I was rolled out of the operating room, still half-anesthetized, holding a wonderful trophy: two beautifully colored endoscopic pictures of the round, green monsters swimming in a tunnel of bile and blood. Saved by high-tech medicine, I could return home that same night and enjoy my first real meal in weeks. I proudly showed off my pictures to anyone who wanted to see them—finally, real evidence for an ailment I had, after all, not imagined!

My personal story of pain and liberation gave me a few valuable insights into the powers of medical imaging—apart from the obvious observation that visualizing technologies play a crucial role in contemporary medicine. Apparently, I had put such trust in the diagnostic visual evidence (gastroscopy and ultrasound) that I was ready to deny my own experience of pain. Thanks to my gastroenterologist, who relied more on anamnesis and the sheer number of blood tests, I am able to tell this story today. Even more paradoxical was my eagerness to show off the visual trophy of this fantastic voyage: a tedious tale of physical discomfort becomes much more exciting when complemented by awesome full-color pictures. The pictures not only "mediated" the narrative of my ordeal, but demanded awe and respect for the heroic rescuers who had captured my little green enemies. I heard myself explain the details of my endoscopic journey over and over again, implicitly advertising the technological advancements of reparative medicine and, in the triumph of relief, seriously understating the preceding pain, suffering, and risk involved in this medical adventure.

These insights form the backbone of this book. Medical images of the interior body have come to dominate our understanding and experience of health and illness at the same time and by the same means as they promote their own primacy. Medical imaging technologies have rendered the body seemingly transparent; we tend to focus on what the machines allow us to see, and forget about their less-visible implications. In order to understand these implications, we need to widen our perspective from a singular medical view to the cultural context in which imaging technologies have evolved over the past centuries. In both contemporary and historical medicine, the development of medical imaging technologies has been intimately tied in with the instruments that enable us to see our own bodily interiors and, simultaneously, witness the technological advancements of medical science.

ACKNOWLEDGMENTS

Like most books, this one is a product of collaboration, collegiality, and friendship. What started out as a concept gradually germinated into a plan and finally materialized into a research project of which this book is only one of many offspring. Long before "The Mediated Body" became an official project at the University of Maastricht, generously funded by the Netherlands Organization for Scientific Research, my colleagues Rob Zwijnenberg, Renée van der Vall, Jo Wachelder, and Bernike Pasveer shared my enthusiasm for this topic, and their inspiration has been indispensable to the completion of this book. In recent years, the "mediated bodies" (or simply "the bodies," as our team is also nicknamed) have been joined by Maud Radstake, Jenny Slatman, Mineke te Hennepe, and Rina Knoepf, whose valuable contributions are likely to turn the project into a success.

The Department of Arts and Culture at the University of Maastricht allowed me precious research time to work on this book, and has been supportive of my work even after I decided to transfer to the University of Amsterdam. I would like to thank Wiel Kusters, Rein de Wilde, Wiebe Bijker, and Karin Bijsterveld for their support. The many students who have, over the past five years, participated in my seminar on body visualization will find in this book remnants of discussions that floated around during our lively meetings. Martijn Hendriks was a great help in finding the right illustrations.

Over the past five years, friendly colleagues on both sides of the Atlantic have read earlier versions of chapters and have in various ways contributed to the many building blocks that make up this book. I would like to thank Jan Baetens, Jay Bolter, Lisa Cartwright, Hugh Crawford, Joe Dumit, Richard Grusin, David Keevil, Christina Lammer, Thierry Lefèbvre, Catrien Santing, and Ginette Verstraete for their input and suggestions. Monica Azzolini offered valuable information on Renaissance anatomy.

The Netherlands Organization for Scientific Research (NWO) has supported my work in many ways. Without their travel grant, I could not have pursued this project; their substantial financial support to our Mediated Body program has given an intellectual boost to interdisciplinary research in the humanities and social sciences. Two American universities hosted me as a visiting scholar while writing this book: the Georgia Institute of Technology in Atlanta and the Massachusetts Institute of Technology in Cambridge. Both offered a stimulating intellectual environment from which I have substantially profited. Three archives, the Dutch Film Museum, the Dutch National Audiovisual Archive, and the George Eastman Archive at the University of Rochester, New York, were instrumental in finding the right sources. I would like to thank the many librarians and archivists who have facilitated my search. With the help of doctors and technicians from the Academic Hospital in Maastricht, my very limited technical knowledge in matters of medical imaging was somewhat enlightened; their patience and enthusiasm is greatly appreciated.

Some chapters in this book have roots in Dutch or English publications. I have used parts of the following previously published materials: "Digital Cadavers: The Visible Human Project as Anatomical Theater," *Studies in the History and Philosophy of the Biomedical Sciences* 31 (2000): 271–85; "Bodyworlds: The Art of Plastinated Cadavers," *Configurations* 9 (2001): 99–126; "Bodies without Borders: The Endoscopic Gaze," *International Journal for Cultural Studies* 4 (2001): 219–37; and "Medical Documentary: Conjoined Twins as a Mediated Spectacle," *Media, Culture and Society* 24 (2002): 37–56. Versions of some chapters have appeared in Dutch in *Het Transparante Lichaam: Medische Visualisering in Media en Cultuur* (Amsterdam: Amsterdam University Press, 2001). I would like to thank the various contributors of illustrations for giving permission to publish them in this book.

The University of Washington Press made the process of manuscript preparation and editing a total delight. Thanks to Phillip Thurtle and Robert Mitchell for initiating the In Vivo series and for extending their enthusiasm to my book, and to Jacqueline Ettinger for her kind and insightful editorial support. Kerrie Maynes proved to be a superb copyeditor.

Ton Brouwers, as always, has been my sharpest critic and most unrelenting supporter; without his eminent editorial skills and loving friendship, this book would not be what it has become.

THE TRANSPARENT BODY

CHAPTER 1

MEDIATED BODIES AND THE IDEAL OF TRANSPARENCY

THE TRANSPARENT BODY IS A CULTURAL CONSTRUCT MEDIATED BY MEDICAL INSTRUMENTS, MEDIA TECHNOLOGIES, ARTISTIC CONVENTIONS, AND SOCIAL NORMS. IN THE PAST FIVE CENTURIES, A HOST OF TECHNICAL TOOLS HAVE BEEN USED TO VISUALIZE THE INTERIOR BODY. BUT HAS THE BODY, AS A RESULT, BECOME MORE TRANSPARENT? TRANSPARENCY, IN THIS CONTEXT, IS A CONTRADICTORY AND LAYERED CONCEPT. IMAGING TECHNOLOGIES CLAIM TO MAKE THE BODY TRANSPARENT, YET THEIR UBIQUITOUS USE RENDERS the interior body more technologically complex. The more we see through various camera lenses, the more complicated the visual information becomes. Medical imaging technologies yield new clinical insights, but these insights often confront people with more (or more agonizing) dilemmas. Behind the alluring images hide ethical choices, and medical interventions are often stipulated by artistic inventions. The mediated body is everything but transparent; it is precisely this complexity and stratification that makes it a contested cultural object.

Between the early fifteenth and the early twenty-first century, a plethora of visual and representational instruments have been developed to help obtain new views on, and convey new insights into, human physiology. From the pen of the anatomical illustrator to the surgeon's advanced endoscopic techniques, instruments of visualization and observation have mediated our perception of the interior body through an intricate mixture of scientific investigation, artistic observation, and public understanding. Each new visualizing technology has promised to further disclose the body's insides to medical experts, and to provide a better grasp of the interior landscape to laypersons. The mediation of human bodies, both historically and contemporarily, has occurred primarily by way of two types of technologies: medical imaging and media technologies.

Ever since the Renaissance, looking into the body's interior has constituted the empirical imperative of medical science. Physicians and scientists gained knowledge about health and disease mainly through dissection and close inspection of cadavers. The emergence of modern imaging technologies coincided with the introduction of a new medical gaze: in 1895 Wilhelm Röntgen became the first to discover a technique for inspecting the living interior body without having to cut it open. Little more than a century after this revolutionary invention, the human body has become highly accessible and penetrable by optical and digital tools. Corporeal transparency is thus primarily a consequence of an increasing number of sophisticated medical imaging technologies, enabling the doctor's eye to peer into the human body.

Apart from medical tools, media technologies have also substantially contributed to the body's transparency. Mass media show us images of the tiniest and most private aspects of the human interior. Documentaries on viruses, photos of a fertilized egg, moving pictures of a fetus in the womb, films of complex neurological operations—is there anything left to hide from view? In recent decades, the body has acquired a pervasive cultural presence, fully accessible not only to the doctor's professional gaze, but also to the public eye. Indeed, impressive medical imaging technologies have enabled this new transparency, but the mass media, engaged in an equally successful effort at permeating our social and cultural body, gratefully pay lip service to the eagerness of doctors and technicians to bring their ingenuity into the limelight. The media's insatiable appetite for visuals has undoubtedly propelled the high visibility of the interior body in modern-day culture.

Mediated bodies are intricately interlinked with the ideal of transparency. Historically, this ideal reflected notions of rationality and scientific progress; more recently, transparency has come to connote perfectibility, modifiability, and control over human physiology. The ideal of transparency is not simply pushed and promoted by medical science. The transparent body is a complex product of our culture—a culture that capitalizes on perfectibility and malleability. In our contemporary world, the interrelations between medicine, media, and technology are all but perspicuous.[1] After some preliminary remarks about the role and function of medical imaging techniques, I will look closely at their relation to media technologies, and end with an elucidation of the transparent body as a social and cultural construct.

MEDICAL IMAGING TECHNOLOGIES

Before the discovery of X rays, doctors depended primarily on their senses (sight, touch, hearing) in order to imagine the interior body. Direct sensory perceptions are still important diagnostic means for physicians, even though they depend increasingly on the optical-mechanical eye.[2] Since the nineteenth century, doctors have used mechanical instruments to translate bodily movements or sounds into readable, visual graphics; the French engineer Etienne-Jules Marey, for instance, invented the electrocardiogram (ECG) and a host of other inscription devices.[3] Marey also systematically deployed photography to minutely register the movements of limbs and muscles in order to obtain a rudimentary knowledge of human kinetics.

Although a number of visualizing techniques have their roots in eighteenth- and nineteenth-century optics or mechanics, the discovery of *X rays* ushered in the era of modern imaging technologies. Since then, we have witnessed the introduction of numerous other techniques. *Ultrasound,* a visual diagnostic practice based on the physics of sound, has gradually become a routine screening instrument for fetuses. The *endoscope,* featuring a mini-camera attached to a flexible cable, is inserted into the body via a tube and sends video signals to a monitor in the operating theater. *Computed tomography* (CT) utilizes X rays to produce ultrathin cross sections of the body; a large number of digital cross sections can be recombined to form three-dimensional representations—for instance, of organs. *Magnetic resonance imaging* (MRI) produces similar slices, but uses magnetic fields, rather than X rays, to penetrate even bone material. *Positron emission photography* (PET) is based on the use of radioactive isotopes, which, when injected into the patient, allow the researcher to study brain functions *in vivo.* The *electron microscope* (EM) gives visual access to the tiniest organic units, such as molecules, which can be magnified up to half a million times.[4]

The development of medical visualizing instruments is commonly viewed as a technological evolution, occasionally accelerated by revolutionary leaps; technicians point to digitization as the latest optical-information revolution.[5] Sociologists and historians of technology have produced quite a few case histories of single imaging instruments, relating their invention and innovation to emerging professional specialties, such as gynecology or radiology, or to specific industrial contexts.[6] Most scholars restrict their attention to the medical domain, focusing primarily on the instrument's technological refinement or its implementation in medical practice. By contrast, some historians of science account for the way in which medical images become part of the texture of modern life. Yet if they do, they often assume a self-evident, causal relationship: medicine develops instruments such as X-ray machines and endoscopes, after which the resulting images are disseminated in other domains, such as art, politics, or popular culture. In her history of medical imaging techniques, Bettyann Holtzmann-Kevles exemplifies this approach by tracing "the technological developments and their consequences in medicine" before turning to "the impact that this new way of seeing had upon society at large."[7]

It is widely assumed that medical imaging technologies reveal the body's interior in a realistic, photographic manner and that each new instrument produces sharper and better pictures of latent pathologies beneath the skin. Traces of this belief resonate in Holtzmann-Kevles's metaphorical claim that as technology improved, "physicians gradually pushed back the veil in front of the internal [body]."[8] Although there is an obvious kernel of truth in this way of thinking, it also reduces a host of complicated, multidirectional processes to a single straight-arrow story of technological progress. Of course, every new medical imaging apparatus provides more knowledge about health and illness, but the same technologies do much more: they actually affect our view of the body, the way we look upon disease and cure. Holtzmann-Kevles's assertion typifies the Western ideal of fully transparent and knowable bodies. The myth of total transparency generally rests on two underlying assumptions: the idea that seeing is curing and the idea that peering into the body is an innocent activity, which has no consequences. Popular media reflect and construct this myth, the ideological underpinnings of which are deconstructed below.

Common belief in the progress of medical science relies in part on unswerving confidence in the mechanical-medical eye: that better imaging instruments automatically lead to more knowledge, resulting in more cures. From visualizing to diagnosing seems a minor step—a doctor just needs to "see" in order to find a remedy. Every newly developed technique appears to lift the veil of yet another secret of human physiology. If we combine all computer-generated images into one comprehensive scan—the digital integration of CT, X ray, electron microscopy, PET, and MRI—might we ultimately be able to "map" each individual body? It is a truism that X rays have been a crucial element in the diagnosis, prevention, and cure of tuberculosis; it is equally common knowledge that ultrasound technology has enabled doctors to recognize fetal defects at an early stage of pregnancy. However, not every disease or aberration is visible or "visualizable." The idea that, by combining all imaging technologies, we can create an ultimate map of a human body is as presumptuous as the claim that we can find the meaning of life by mapping the human genome. And yet, patients often blindly trust the panoptic nature of the mechanical-clinical eye.

Despite the equation of *seeing* and *curing* in popular media, better pictures do not automatically imply a solution. Medical scans often show irregularities or abnormalities, the progression of which doctors cannot predict, or for which there is no cure. Innovation in medical imaging technologies is the result of a constant attunement of machines and bodies, of procedures and images, of interpretations and protocols.[9] We can never assume a one-to-one relationship between image and pathology: looking at a scan, medical experts may identify signs of potential aberrations, but their interpretations are not necessarily univocal. To a certain degree, medical-diagnostic interpretation of a scan is always based on a consensus between specialists; it may take years before consensus transforms into a reliable heuristic protocol, and even after applying a technique for several decades, its images may still give rise to different interpretations.[10] Reading X rays, endoscopic videos, or MRI scans involves highly specialized skills that require substantial training and practice. In addition, with each new instrument or innovation, doctors have to readjust their reading and interpretation skills. While advanced machines render our bodily interior seemingly more transparent all the time, the images they produce hardly simplify our universe.[11] Seeing often leads to difficult choices, multifarious scenarios, and thus complicated moral dilemmas.

The other important implicit assumption involving imaging technologies is the belief that looking into the body is an innocent activity. This belief allows for reasoning such as "we can always take a look and if we don't see anything, nothing happens," or "bodies remain untainted if we only touch them with our gaze." Philosophers and sociologists of science have already sufficiently countered this axiom.[12] Ian Hacking, for instance, argues that every look into a human interior is also a transformation—"seeing is intervening"—because it affects our conceptualization and representation of the body.[13] Medical imaging technologies not only shape our individual perceptions, but also indirectly contribute to our collective view on disease and therapeutic intervention. The definition and acknowledgment of a disease often depend on the ability of medical machines to provide objective visual evidence, and insurance companies may not cover diseases unless they are visually substantiated.[14]

Relying on the mechanical-clinical eye has direct and indirect consequences: it directly influences a patient's medical treatment and indirectly structures healthcare policies. For instance, more advanced ultrasound machines show more fetal defects at an earlier stage of pregnancy; the technical ability to detect rare fetal abnormalities becomes the technical imperative to offer such scans to all pregnant women. Mapping the human genome, far from being an "innocent" exercise in charting all possible genetic sequences, is bound to affect future viability decisions (and insurance policies) about whether a fetus's genetic vitals warrant gestation and birth. New imaging techniques are often initially deployed as individual diagnostic tools before becoming screening instruments; in the process, they contribute to the creation of risk groups. Looking into a body and mapping its organic details is never an innocent act; a scan may confront people with ambiguous information, haunting dilemmas, or uncomfortable choices. This predicament, including its ethical, legal, and social implications, does not simply arise as a *consequence* of new medical imaging technologies, but it is *intrinsic* to their very development and implementation.

It goes without saying that new medical imaging technologies have greatly advanced medical diagnostics and research; sophisticated tools help medical professionals to detect disorders at a much earlier stage, or they assist them when planning intricate operations. In this book, I intend neither to hail the triumphs of modern medicine nor to detract from its achievements. My aim is to provide a cultural analysis of medical imaging, to unfold the cultural complexities involved in medical imaging instruments and products as well as their uses and meanings, both inside and outside medicine.[15] Every year, approximately 250 million scans are made in American hospitals alone.[16] According to a report by the Blue Cross and Blue Shield Association, diagnostic imaging is approaching a $100-billion-a-year business in 2004, about a 40 percent increase since just 2000.[17] Most people view the ultrasound scanner, endoscope, or CT scanner as medical-technological appliances and consider the images they produce expendable, their significance beyond the walls of the clinic being close to zero. Yet, in recent decades, these machines and images have rapidly become an integral part of our visual culture. Medical imaging technologies have attained a prominent cultural presence in their own right, but there are significant overlaps with other technologies and cultural processes—in particular, those involving the interests and values of the mass media.

MEDICAL IMAGES AND MEDIA TECHNOLOGIES

Visualizing instruments used for medical diagnostics are related to media technologies on at least three levels. First, their technological developments tend to go hand in hand, meaning that innovations in one domain benefit technical advancements in the other domain. The invention of X-ray films in the 1950s, for instance, designed to record patients' lung movements, was made possible by the invention of the image intensifier; although the medical application never caught on, the image intensifier gave a boost to the production of television. Endoscopy's various stages of development have been closely connected to advances in media technology, such as (color) photography in the 1960s and television and video technology in the 1980s. The imagery produced by the mini-camera a surgeon inserts into the body of a patient ends up on a television screen. Cardiovascular scanners employ a technique that is used in compact-disc players, while MRI and CT scanning would be impossible without advanced computers. Digitization in general has caused medical and media instruments to merge. In the future, image processing, management, communication, and analysis will all coalesce in a computer-mediated system, further reducing the distinctions between media and medical technologies.

Secondly, in addition to their technological coevolution, media soon began to function as an intermediary for medical knowledge. After Wilhelm Röntgen discovered X rays in 1895, the technology gradually became standard in clinical settings. Similarly, the invention of film in that same year caused the burgeoning of cinema as a popular attraction at fairs and traveling shows. These two developments intersected in the first decades of the twentieth century, as X-ray images appeared on big cinema screens in tuberculosis-prevention campaigns.[18] This new way of disseminating interior-body imagery among the public at large has undoubtedly contributed to a rising public interest in medical issues; large-scale displays of pictures revealing hitherto unseen mysteries generated excitement and were considered attractive and aesthetically pleasing by many. It is no coincidence that today we are still bombarded with images of fetuses, beautifully colored PET scans, or black-and-white shadows of ultrasound pictures; the abundant use of medical imagery in newspapers, movies, television, and magazines suggests that these media ventures are all eager to cash in on this phenomenon.

One could safely assume that the visualizing trend in medicine was promoted by mass media eager to exploit the power of fascinating, authoritative images. But the opposite is equally true: doctors and hospitals, keen on public relations, recognized the enormous publicity value of intriguing bodily images. In the modern welfare state, health tops the list of public concerns, and, understandably, the media cater to this popular priority. Whether the growing presence of medical images in mass media is the result of more and better medical imaging technologies, or the consequence of the ubiquitous, all-pervasive camera in private affairs, is hard to tell. The "mediation" of medicine is part of a more general trend to allow cameras into our intimate lives.[19] Media's ubiquitous presence in the rituals of our individual, private domains has blurred the boundaries between what used to be separate spheres.

Thirdly, as the above already suggests, medical and media technologies converge in their production of visual spectacle—displaying the inside of a human body. Shortly after the invention of film, doctors started to deploy the camera to record surgical interventions. More than a century later, the presence of cameras in operating rooms hardly raises eyebrows; recently, a delivery of triplets by cesarean section in a Dutch hospital was broadcast live on the Internet.[20] Public interest in medical procedures, however, preceded the introduction of film. The tradition of publicly displaying cut-open bodies dates back as far as the late Middle Ages. In sixteenth- and seventeenth-century Europe, anatomical theaters and public anatomical lessons attracted large crowds; dissected cadavers formed a fascinating spectacle because they were associated with intimacy, sex, and violence. Cutting into a person's body—whether for anatomical or surgical reasons—always attacks that person's physical integrity. A scalpel's incision confronts onlookers with blood, knives, and bare organs, and so does a recording of this procedure. Recent imaging and operating techniques that leave the body's surface intact appear to weaken connotations of sexuality, violence, or spectacle; the endoscopic camera directs our view from inside the body, circumventing the skin, while MRI scans allow us to view cross sections of the body without it having to be dissected at all. Yet these new technologies do not so much eradicate as change the nature of the body as spectacle: spectacle is now a feature of the technology that draws the public eye into the body, enabling the public to see what the surgeon sees.

There are obvious distinctions between cameras deployed for medical reasons and those deployed for media purposes, between the gaze of the surgeon and the gaze of the layperson. The emphasis in this book, however, is on convergences in these areas rather than divergences. Medical and media technologies are both technologies of representation. They provide particular ways of accessing the internal body, and determine its depiction; the resulting representations, in turn, fashion our knowledge of the body and set the parameters of its conceptualization. This recursive process, in which perception, representation, conceptualization, and knowledge formation are inextricably intertwined, has historically involved other professional groups besides doctors. For one thing, the medical profession has relied on illustrators to translate physiological or anatomical insights into comprehensible depictions. Art and medicine, as Canadian media scholar Kim Sawchuk argues, "have worked in tandem in the production of knowledge of our bio-being, not only to produce specific representations, but to develop a particular way of knowing through techniques of visualization."[21] During the Renaissance, pencils and brushes were the prime tools used by the artist to transfer images from the mind's eye to paper. The arrival of mechanical instruments did not dislodge the artist; on the contrary, illustrators still function as important mediators between anatomical insights and their visual representations. Modern medical imaging technologies and computer graphics software are, like the illustrator's pencil, indispensable aids in the production of images. The historical continuities between representation technologies—both medical and media—are paramount to a comprehensive understanding of how medical knowledge is represented or representable. Visual depiction of anatomical data, even today, is defined as much by medical technologies as by artistic traditions and styles.

According to Michel Foucault, our bodies have become "sites where organs and eyes meet."[22] More precisely, the mechanical-clinical gaze—the gaze directed and mediated by imaging technologies—detaches a body from a person, a process that Foucault refers to as "externalizing the internal." It is precisely its dissemination outside of medicine that has popularized the mechanical-clinical gaze. For instance, we attribute different meanings to ultrasound pictures or microscopic images outside a clinical context. Frequent use of X-ray shadows in advertisements or of endoscopic images in motion pictures has not familiarized the audience with their medical interpretations, but has added a variety of connotations to their pictorial styles. Ultrasound pictures are commonly associated with babies or prenatal care, PET scans have already begun to connote "psychic dysfunction" or schizophrenia, and MRI scans automatically elicit mental images of cancerous tumors. In everyday culture, we see so many of these images that we are tempted to believe we understand their (medical) meanings. This load of connotations cannot be disposed of upon entering the hospital for a scan, and the circulation of denotations and connotations makes it hard to tell medical from nonmedical meanings. Consequently, the clinical gaze distributed in culture affects and shapes our collective view of the body and the way it can and should be treated in medicine.[23] Even if this mutual shaping of the (mechanically mediated) gaze and the formation of collective norms and values cannot be verified empirically, this process proliferates in a visual culture that privileges sight and spectacle.

The way doctors visualize pathologies also affects the way society envisions and addresses health issues. Once again, my main concern is not with establishing any causal relationships between processes of visualizing diseases and curing them, but with the role of medical imaging technology in the social and cultural construction of disease. Magnified pictures of eggs and sperm have most likely contributed to our communal concept of infertility as a disease and in vitro fertilization as its remedy.[24] Microscopic enlargements of T cells, frequently appearing in flyers and public-affairs magazines, not only served to define the HIV virus as a dangerous enemy, but also promoted the effectiveness of AZT and other AIDS drugs. Showing a virus in situ is as effective as showing or visualizing the weapons used to fight the intruder.[25] Successes in medicine become evocative narratives in popular culture: medical imaging technology produces images of pastoral bioscapes threatened by external or internal invaders (viruses or tumors). The images and text combined produce the persuasive narrative of a body under siege by foreign armies and protected by the chemically fortified immune system.[26] Such metaphors and images, in turn, foster a particular conceptualization of disease, one that may spur the development of new technologies. Our view of genetic engineering, for instance, is determined in part by our mental images of what genes are and how they function.[27] Images are the products of instruments; but instruments are also the products of our imagination. The significant role of images and imagination in the construction of corporeality is one of the prime motivations for cultural critics to analyze and theorize medical imaging.

But there are more reasons to be interested in this subject. Optical technologies are also techniques of illusion, deception, and voyeurism. When medical and media technologies merge, so do their visual codes, causing a mixture of different modes of looking. Television and movie directors consciously exploit the ambiguity generated by this convergence. When they dramatize and narrativize the clinical images they absorb, viewers must oscillate between the "objective" pictures produced by medical instruments and the "subjective gaze" directed by the television camera. If shots of a surgical operation are part of a documentary, we interpret them differently than we would if they were part of a clinical film or, for that matter, a feature movie.

The increasingly pervasive presence of television cameras has triggered a variety of ethical questions: What are the limits of showing interior bodies, surgical interventions, and other medical images? How does the camera push the limits of privacy, human integrity, and public taste? Contemporary Western societies stress the importance of ethics in issues of predictive and preventive medicine, such as human cloning or assisted reproduction. If we take medical-ethical reflection seriously, we need to give a great deal more prominence to the role of representational technologies in the construction of norms and values. Yet how are we to assess the impact of the hundreds of medical scans appearing in the media every day, whose function varies from almost diagnostic to purely symbolic or aesthetic? A colorful, retouched PET scan on the cover of a professional medical journal serves both an educational and an aesthetic function; an ultrasound scan in the logo of a news item on health care exemplifies a purely symbolic use. The use of medical images wavers between data sharing and entertainment. Frequent exposure to pictures evidently affects our norms and values, even if unconsciously. Watching endoscopic operations on television not only familiarizes viewers with the surgical gaze from within the body, but concurrently redefines public standards on integrity and privacy. Films of breast enlargements or cosmetic facelifts broadcast on public television both reflect and construct contemporary norms concerning the perfectible body. Medical documentaries or talk shows often present the human body as a mechanical-organic entity that can be disassembled and reassembled at will; doctors and surgeons tinker with bodies until they look like the fashion models in magazines. Media images of bodily interiors are often coupled with pictures of beautiful, happy, and healthy people. A transparent interior—medically translucent and endlessly modifiable—seems a sine qua non for a perfect exterior.

Indirectly, or perhaps inadvertently, media that publicize medical-technical issues also affect health-care funding. Much of medical television programming—whether information or entertainment—focuses on high-tech surgical interventions *(The Operation* and *Extreme Makeover),* heroic rescue operations *(Rescue 911),* or doctors' dramatic struggles for the life of a patient rushed into the emergency room *(ER* and *Chicago Hope).* Doctors and surgeons are manifest as gods, the hospital as a sacred institution, and technology as a deus ex machina. As visual spectacles, however, these programs obscure the financial interests of the health sector, commonly represented as charitable "service" institutions rather than capital- and technology-intensive industries. Deployment of medical imaging techniques in popular media places the emphasis on reparative medicine at the expense of palliative or preventive health care. Not that the media never pay any attention to the patients populating the oncology ward or to nurses caring for Alzheimer's patients, but on television, these pale in comparison to technological life-saving excursions. Media attention often translates into public relations, and public relations keep money-devouring high-tech medicine going. In more than one respect, the marriage of media and medical technologies is mutually lucrative. Public appropriation of medical technologies is seldom neutral, and always serves economic or political interests.

Medical programs promoting imaging technologies can be seen from a number of angles and from a variety of viewpoints, depending on who is watching. In contrast to doctors and biomedical scientists who are extensively trained to interpret medical images, the viewers of medical television programs rarely have any professional training in decoding the meanings produced by a combined medical-media apparatus.[28] As science-studies scholar T. Hugh Crawford has convincingly argued, an emphasis on the "ethics of seeing" might significantly affect the framing of medical-ethical issues. Certain medical movies and documentaries, according to Crawford, "underscore the troubling epistemological relations of optical technologies and people—of looking, looking through, and being looked at. Such relations seem transparent . . . but they are instead fraught with complexity and produce by their very transparency an unacknowledged and overdetermined way of viewing and acting in the world."[29] In a culture that increasingly concedes private grounds to public cameras, medical-ethical issues are media-ethical concerns as well; the ethics of representation, therefore, are part and parcel of the aesthetics of display. Aesthetically appealing images fascinating large crowds tend to override legitimate questions of ethical permissibility and educational value.

THE TRANSPARENT BODY AS A CULTURAL CONSTRUCT

Transparency has been a constant ideal in Western medicine, but that ideal has not remained static over the ages. Historically, this ideal connoted notions of rationality and scientific progress; by looking into bodies (mostly corpses), doctors could increasingly understand the secrets of human physiology. In the late nineteenth century, the notion of corporeal transparency, induced by ocular instruments, became associated with ideas of personal and public hygiene. The emergence of modern imaging instruments at the end of the nineteenth century introduced a new notion of transparency. As German media theorist Friedrich Kittler has explained, the twentieth century gave a boost to the use of "inscription technologies," mechanical devices such as the gramophone and the movie camera that produced exact representations of human bodies.[30] In the area of medical imaging, inscription technologies such as X ray, ultrasound, and endoscopy seek to dispose of mediation (such as an artist's drawing) and instead record the interior body directly onto a machine. The mechanical gaze into *living* bodies not only enhanced the body's transparency, but also its manipulability. In the twentieth century, the ideal of transparency has become associated primarily with medical notions of perfectibility or modifiability. Today, artists and media producers continuously confront us with the concept of the transparent body, with minute pictures of bodily details the very existence of which was still a matter of guesswork only several decades ago. With the help of advanced medical imaging techniques, futurologists promise, doctors will soon need just a single hologram scan to identify disease and remove potential physical threats. As we shall see, the ideal of transparency as a precondition for medical power and control over human health and longevity is inextricably tied in with the emergence of these new inscription technologies.

Imaging technologies, even if we call them medical, are never exclusively generated within medicine before they affect other domains—or, for that matter, culture at large. From the earliest stages of their invention, imaging techniques have produced more than medical evidence or visual illustrations. Many medical imaging tools originated outside the medical domain, and much high-tech medicine is closely related to general social and technical developments. Connecting medical and media technologies, I intend to illuminate how the ideal of corporeal transparency is rooted in Western culture. As a normative ideal, it codetermines individual and social norms and values, while, on a policy level, it directs choices in health care.

This book challenges the somewhat simplified notion that new imaging technologies lead to more knowledge and thus lift the veil from the interior body. Several cultural theorists and anthropologists have questioned a priori distinctions between medical and nonmedical imaging technologies, thereby disputing the assumed teleology of medical technologies and their social or cultural impact. I subscribe to the approach taken by Lisa Cartwright, who equally considers medical and media technologies as representational tools, producing meanings at a particular historical moment. In terms of methodology, I favor sociologist Jackie Stacey's preference for analyzing medical practices through which the cultural meanings of technologies are constructed. However, I find these approaches limited when it comes to understanding more about actual implementations of imaging technologies in contemporary medical-cultural practices.[31] Therefore, I have found it particularly useful to also incorporate strategies advocated by (medical) anthropologists and sociologists of technology, such as Joseph Dumit, Anne Beaulieu, and Emily Martin.[32] Moving from the surface of the body to its virtual interior, each of the following chapters will analyze a particular imaging technology that can be considered as cultural rather than strictly medical. Throughout the chapters, I will emphasize continuities between historical and contemporary ways of imaging and imagining the interior body by addressing three types of issues situated at the crossroads of medicine and media. First, there are questions concerning the responsibility of images—who is responsible for representation of the interior body? Second, what are the interests of medical professionals and media producers—particularly if these converge? Third, what is the role of bodily representations in the formation and shifting of norms and values concerning perfectibility and modifiability, but also concerning privacy and bodily integrity?

Chapter 2 looks at "spectacles of nature" in the form of conjoined twins, as they were (and still are) observed from the outside by curious audiences. Although the popularity of the freak show declined in the early twentieth century, due to the changing mentality and growing medicalization of Western society, it never disappeared completely, but metamorphosed into a medicalized-mediated spectacle. Since the invention of the movie camera, the separation of conjoined twins has been the subject of medical documentaries, from the films of Dr. Doyen in early-twentieth-century France to recent American public broadcast documentaries. Using Michel Foucault's and Guy Debord's theories, I will analyze the cinematic and contextual changes in the co-production of the medical-media spectacle.

The questions of responsibility for the representation and exhibition of interior bodies informs chapter 3, which focuses on the work of the German anatomist Gunther von Hagens. This anatomist-artist developed a preservation technique called "plastination," by which a chemically treated corpse is modeled into a sculpture. Fossilized interior bodies were put on display during Bodyworlds, a series of exhibitions of plastinated cadavers that traveled all over Europe, attracted millions of visitors, and caused a major public outcry. The anatomy-as-art controversy is examined from a combined medical-historical and art-historical perspective. The histories of anatomy and art provide a necessary interdisciplinary subtext for probing some of the current dilemmas in the evaluation of postmodern anatomical practices: to what extent do we accept dead bodies as art and who decides upon the boundaries between anatomy and art?

In constructing an ideal of transparency, the interests of medical professionals and media producers often converge; chapter 4 explores how this convergence is epitomized by a particular imaging technology used to visualize the inner body by entering it through natural or created orifices: the endoscope. Sketching the past, present, and future of endoscopy, I advance the broader argument that medical technologies are the material embodiment of collective desires and fantasies, which, at the same time, spur the very design of these technologies. Using the Hollywood classic *Fantastic Voyage* as a departure point, I argue that the body voyage remains a central trope in medical documentaries. The myths of transparency and nonintervention have prevailed in both the production and popular dissemination of the endoscopic gaze.

Historically, the introduction of each new imaging technology signaled the emergence of a new visual regime that was never simply restricted to the medical domain. Chapter 5 shows how in the early 1900s the X ray was hailed as a new instrument of objective verification and indisputable proof of tuberculosis, a widespread disease at that time. But in addition to being considered a new medical imaging technology that could bring to light previously invisible parts of the living body, X rays were believed to be a sort of superphotography that could prove the existence of immaterial substances, the materiality of things heretofore unseen. Thomas Mann's *Magic Mountain* shows how X rays were also thought to visualize intimate feelings such as love, and prove the existence of the spiritual self after death. Mann's novel does not simply reflect these beliefs, but problematizes the cultural conceptions inspired by medical-scientific axioms.

Medical imaging technologies play a constitutive role in the formation of norms concerning the perfectibility and modifiability of the human body, and the representational value of medical images forms a locus of contestation. Chapter 6 highlights the role of ultrasound pictures in the display of the "invisible" fetus growing in a living body. Ultrasound allows doctors to detect pathological growth in the early stages of pregnancy. Beyond the boundaries of the clinic, however, the sonogram has taken on a variety of cultural meanings, most of all as "baby's first picture." This chapter deals with the rise and fall of a peculiar phenomenon in the Netherlands between 1985 and 2000: the "ultrasound-for-fun clinic." Parallel to the growing use of ultrasound in birth clinics, a private market emerged for "fetus photography." The ensuing struggle between gynecologists, midwives, and unlicensed sonographers for its regulation is in fact a contest between the cultural and medical meanings of ultrasound. Why and how do medical professionals want to reinstate the boundary between medicine and culture that they had so strategically trespassed before?

The contested boundary between "real" bodies and mechanical-medical representations is erased even more subtly by digital imaging technologies. Chapter 7 provides a cultural-historical analysis of the Visible Human Project (VHP), a large-scale scientific program funded by the American government that aims at producing two "standard" digital anatomical bodies, developed from cadavers, to be used by medical students and researchers all over the world. Technologies such as MRI and CT scanning are crucial to the "dissection" of a digitized human body, whose data have been reconfigured and made accessible for medical purposes through the Internet and for a general audience through popular CD-ROMs. Pivotal to understanding the aims and content of this contemporary digital dissection project is an understanding of anatomical theaters and public dissections in the Renaissance. Besides touching on issues of privacy and bodily integrity, the VHP raises poignant normative questions concerning education, crime and punishment, and entertainment.

While some chapters in this book focus on the cultural history of specific imaging technologies, others capitalize on the role of media and art in the dissemination of the mechanical-clinical gaze; again others dwell on medical instruments serving as media technologies. Each individual chapter comprises a self-contained narrative elucidating the transparent body as a contested concept, using the continuity between historical and contemporary ways of looking at and representing the body as a point of departure. Applying cultural analysis to a number of peculiar phenomena in the area of medical imaging unveils the many evident, though sometimes subtle, intersections between medicine and culture. The relevance of this type of analysis is at least partially validated if patients going in for scans, medical and media professionals, or those who merely enjoy watching surgical procedures and medical dramas on television, after reading this book, look differently at the images, instruments, and practices involved in body imaging.

CHAPTER 2

THE OPERATION FILM AS A MEDIATED FREAK SHOW

TAKING A STROLL THROUGH THE PARK ON A SUNNY AFTERNOON, I CAN HEAR FROM A DISTANCE SOMEONE PLAYING OLD DYLAN TUNES. IT TAKES A WHILE FOR ME TO NOTICE THAT THE YOUNG STREET MUSICIAN SITTING ON A FOLDING CHAIR ATTRACTS QUITE A CROWD FOR SUCH AN ALL-TOO-FAMILIAR ACT. WHEN I APPROACH THE SCENE, I UNDERSTAND WHY: HIS GUITAR IS ON THE GRASS IN FRONT OF HIM AND THE LIMBER TOES OF HIS FEET ARE PRODUCING THE CHORDS. AS BYSTANDERS GENEROUSLY PUT COINS AND DOLLAR bills into the artist's hat, many, like me, are looking searchingly at his upper body. The musician is wearing an old, oversized woolen sweater, sleeves folded inward, his garment mysteriously hiding the exact shape of his torso. When I continue my stroll, I ask myself if it really makes a difference whether or not the musician has arms. Does his physical condition detract anything from the act he is performing, or, contrarily, does it add anything? Either way, I feel slightly uncomfortable at the idea that this young man has to exploit or fake a physical handicap in order to gain people's attention and thus their money.

Several weeks later, while thoughtlessly zapping, I hit upon a television program featuring a seriously obese man, shown in his daily struggle to carry around his 350 pounds of body weight. A medical specialist explains in great detail the inevitable operation this man must undergo if he is to reduce his fat tissue to livable proportions. The ensuing images of a liposuction are not exactly aesthetically pleasing, but my eyes are glued to the screen. After the operation, the man and his doctor enthusiastically discuss its positive outcome, and the patient proudly shows off his new bodily contours. I am having trouble defining the genre of the program: Is this medical information or is it plain entertainment? Does the man warrant my attention because of his serious obesity, or is the plastic surgeon trying to sell me his advanced surgical techniques? I wonder if I should hold the program's creators responsible for my uneasiness, or if I (and millions of other viewers) am to blame for legitimizing this spectacle by watching it.

In our technologically advanced Western society, the phenomenon of individuals publicly exposing their physical handicaps to make money is considered either a regrettable anomaly or a dubious anachronism. Indeed, for centuries, people with physical abnormalities populated the stages of fairs and circus arenas. Exceptionally large or small individuals (giants or dwarfs), extremely fat or tall men or women, persons without a clear-cut gender (hermaphrodites), and people with conjoined bodies (Siamese twins) have always been the subjects of public fascination.[1] Until the early decades of the twentieth century, so-called freak shows were a regular component of road shows and circuses, attracting thousands of visitors who paid money to stare at a live spectacle we would now refer to as a "handicapped person."

Have freak shows, at the turn of the millennium, been relegated to the realm of history? I do not think so. The difference between the freak shows of the nineteenth century and those in our time is that it is no longer the fat man himself who draws a large audience, but the filmed operation on the fat man. We would be embarrassed to stare at the physically challenged exposing themselves on a stage, but we eagerly watch televised recordings of their salvation by medical professionals. This chapter traces how individuals with a particular and rare congenital defect—conjoined twins—have historically been put on display. In the late nineteenth and early twentieth centuries, fairs, circuses, and road shows—vehicles of mass entertainment—provided the immediate context for the public exposure of conjoined twins. Throughout the twentieth century, the abnormality of conjoined twins continued to be a focus of cultural awe, yet the mode of their display shifted to a radically different, mediated domain, namely that of film and (later on) television. Moreover, this change of medium was accompanied by a change of professional domain. If, during the nineteenth century, conjoined twins functioned in the realm of entertainment largely by virtue of their coalescence, in the twentieth century their public presence was by and large due to their surgical separation, as operations became the subject of medical films and television documentaries. Three examples taken from the history of conjoined-twin operation films demonstrate how, when it comes to representing human deformity, twentieth-century spectacle is firmly rooted in the nineteenth-century freak show. The freak show never really disappeared, but took on a new cloak, evolving into the medical documentary, whose appeal is based, to a large extent, on the convergence of medical and media techniques. The most recent, mediated version of the freak show involves a hybrid spectacle in which information, entertainment, public relations, and ideology have fused beyond recognition. It is this subtle interplay of mediation, medicalization, technology, and commerce that causes my discomfort when watching programs like these, knowing that they are made possible by viewers like us.

CONJOINED TWINS AS FREAKS

"The freak is an object of simultaneous horror and fascination because . . . the freak is an ambiguous being whose existence imperils categories, and oppositions dominant in social life."[2] According to American cultural theorist Elizabeth Grosz, conjoined or "Siamese" twins were popular attractions at nineteenth-century fairs (as were hermaphrodites, giants, dwarfs, and hairy "ape men") because their physiology transgressed the boundaries of what was considered ordinary for a human being. Were they two people with one body, or one person with two bodies? On account of a rare deformity triggered during the embryonic stage, monozygotic twins become connected in some concrete fashion—it could be just a matter of shared skin tissue, or their coalescence could involve one or more organs and even limbs. Their appearance invariably undermines the category of the "individual," particularly if the twins share a third leg or are joined at the head. During the Middle Ages and the Renaissance, persons with such abnormalities were categorically dismissed as monsters; stories in which they figured were generally understood in mythical or religious terms.[3] Michel Foucault has extensively argued that, in the course of the eighteenth and especially the nineteenth century, medical explanations increasingly accounted for all sorts of deviations in appearance and behavior.[4] However, it was not until the end of the nineteenth century that the various embryological, genetic, and histological abnormalities were mapped systematically.[5] Yet the medicalization of their condition did not automatically lead to the twins' emancipation from their roles in popular entertainment.

In both Europe and North America, freak shows were standard features of circuses and traveling shows, reaching their heyday in the second half of the nineteenth century.[6] Individuals with rare congenital deformities, deserted by relatives or ousted from their communities, were often forced to surrender themselves to circus managers who commercially exploited their physical deformations.[7] Some "freaks" were literally owned by an agency, either as slaves or (after the abolition of slavery in the United States) as a result of signing a strangling contract that practically reduced them to slaves. Freaks were regularly imported from Asian or African countries; their "exotic" nature functioned as an integral part of the show and was highlighted in promotional campaigns. Physiological abnormality was thus linked with racial or ethnic otherness and defined against the Western standards of normality.[8]

In order to attract as many people as possible, freak-show managers used various methods to underscore the great scientific value of their attraction. Outside the circus tent, the announcer (frequently dressed as a doctor) emphasized the uniqueness of the medical curiosity on display, commonly invoking the authority of specialists who had examined the freak and declared him or her to be a rare exemplar. The freak's performance was accompanied by "expert commentary" from a scientist in a lab coat who explained the phenomenon to a lay audience.[9] After the show, the audience could buy autographed pictures to share with friends and relatives. The sale of photos not only yielded additional income but also attracted new visitors who wanted to see the oddity with their own eyes. Historical sources indicate that freak shows were conspicuously framed as educational events in order to distance them from the mass entertainment commonly associated with fairs and circuses.[10]

Although the public display of conjoined twins antedates the nineteenth century, the term "Siamese twins" is directly tied to one legendary pair: Eng and Chang

Nok, born in Thailand (Siam) in 1811.[11] Eng and Chang were literally inseparable, sharing the skin that held their torsos together, but were also emotionally attached to each other. An American trader paid their mother a substantial sum to take her sons to the United States and show them as an attraction at fairs. None of the many specialists who examined Chang-Eng (the twins preferred this double name) during their lifetimes could ascertain whether or not the brothers had separate livers. Over the years, their managers consulted several famous medical specialists in both Europe and America, yet their main intention was to raise the scientific status of their attraction; evidently, they had not the slightest interest in separating their valuable merchandise. Chang-Eng refused to be separated during the first forty years of their lives.[12] After redeeming themselves from their managers in 1833, they toured around Europe and America for years, until they had earned enough money to purchase an estate in North Carolina. Having acquired American citizenship, they changed their surname to Bunker. Chang-Eng married two sisters, Sally and Adelaide Yates, and the two couples were blessed with a total of twenty-two children. The brothers died in 1874; Chang died of heart failure, after which Eng lived on for two more hours before dying as a result of shock. In due time, the Bunker twins' nickname became a label: the term "Siamese twins" (just like the word "Mongoloid" for children with Down's syndrome) is troublesome for its conjunction of abnormality and exoticism.

Fig. 2. Chang-Eng Bunker, Siamese Twins.

After 1900, public interest in freak shows began to diminish. American sociologist Robert Bogdan accounts for this fading appeal by pointing to the growing medicalization of society during the first decades of the twentieth century.[13] Rather than being accepted as facts of life, congenital deformities were increasingly looked upon as handicaps that could be alleviated or cured through medical intervention. Medicalization, a term theorized by Foucault in more detail, caused a change in public perception to the extent that freaks were no longer regarded as eerie monstrosities but as unfortunate individuals in need of medical help. The growth of medical knowledge—most notably the invention of various visualizing techniques—resulted in a substantial increase in the number of successful operations aimed at separating conjoined twins. X rays helped surgeons to locate and identify organs; diagnostic techniques allowed them to assess whether an operation should be performed at all and, if so, how they should proceed. Starting in the early twentieth century, we can observe the growing dominance of the medical imperative: conjoined twins are no longer doomed to a life of forced corporeal fusion but can be liberated from their predicament through sophisticated surgical intervention. However, I do not agree with Bogdan's view that medicalization caused the disappearance of the freak show; rather, I would argue that medicalization changed the freak show's character. In Foucault's terms, medicalization of the freak gave rise to the normalization of surgical interference in the freak's body. The medical profession's effort to "save" the freak, rather than the freak himself, became the center of attention, projected on public screens.

The vanishing of the "live" freak show coincided with the rise of cinema after 1895. In the early decades of the twentieth century, fairs and road theaters cashed in on movies—the new and undreamt-of phenomenon of bodies moving on a screen.[14] Movie directors selected topics that had always fascinated the public; monstrous creatures frequently appeared in early movies, and the horror movie reached the zenith of its popularity in the 1930s.[15] The potential of film was also explored for medical purposes. In the nineteenth century, photography had proven to be a valuable means of providing visible evidence of physical disorders. Historical photo collections of patients with uncommon pathologies reveal that the *camera medica* produced a hybrid genre, part scientific record of an anomaly, part artistic representation of a deformed body.[16] Due to this ambiguity, these photos could be viewed with either a medical or a voyeuristic eye. This same double layering is found in medical documentaries, a genre that grew more sophisticated as it evolved in the twentieth century.

From the outset, medical professionals deployed the new medium of film to record medical procedures as well as their results. Film was especially popular for recording surgical interventions. Initially, individual surgeons initiated the recording of an operation, but after 1945, hospitals and professional medical organizations took charge of producing these films. Between 1950 and 1970, film gave way to television as the preferred medium, and medical documentaries were increasingly sponsored by the pharmaceutical industry.[17] Since the 1950s, public and commercial broadcasters have become involved in their production and distribution.

Film recordings of the surgical separation of conjoined twins demonstrate how, in the twentieth century, the freak show became a mediated event. French philosopher Guy Debord called contemporary Western culture a "society of the spectacle," implying that all modes of knowledge are subject to the constraints of electronic mediation.[18] The camera is a weapon in the struggle for knowledge and truth as constructed and disseminated by the media apparatus. An important characteristic of the mediated spectacle, Debord suggests, is that various formats and genres—such as information, entertainment, and public relations—have coalesced. The operation film served at least four different goals: First, it was used as a tool for training specialists; celluloid recordings proved a valuable means for familiarizing future surgeons with the fine details of specific surgical interventions, especially in the case of rare operations. Second, the operation film functioned as a verification device. By filming before, during, and after an operation, its results could be visually showcased to outsiders. Third, the operation film served to inform or entertain a lay audience. And, finally, the film had a promotional function; it was produced to popularize medical expertise or technology and to impress viewers with remarkable examples of surgical craftsmanship. Medical films that promote the separation of conjoined twins have rendered the four functions listed above inseparable.

This genre is also subjected to the laws of narrative cinema and visual drama.[19] A surgical separation of conjoined twins is a technically challenging and hence exciting procedure. Its cinematographic inscription, however, entails various elements that contribute to its dramatic appeal: the deviant physiology of the patients involved the ingenuity of the surgeon, advanced medical equipment, and special film techniques. Though arguably the least conspicuous elements in a filmed operation, filming techniques (editing, shots, camera angle) and the conditions of production (setting, funding, distribution) may determine the ultimate shape of this characteristic twentieth-century spectacle. If, in the nineteenth century, conjoined twins figured prominently as popular attractions in live entertainment, in the course of the twentieth century their cultural role has changed radically. They have become part of a mediated spectacle in which the medical specialist takes center stage as the cultural hero who seeks to liberate the twins from their physical confinement.

1902: DR. DOYEN AND THE NEIK SISTERS

In 1898, several years after the Lumière brothers introduced their spectacular invention at public movie screenings in Paris, Dr. Eugène-Louis Doyen (1859–1916) introduced one of the first medical applications of the new medium.[20] A renowned surgeon and owner of a private clinic in Paris, Doyen had already taken a keen interest in the medical possibilities of photography, and the potential of cinema seemed even greater. He hired two *opérateurs cinematographes,* Clement Maurice and Ambroise-François Parnaland, who were willing to collaborate with him on an ambitious plan. Between 1898 and 1906, Doyen recorded some sixty films, each featuring himself performing an operation, occasionally with the help of colleagues.[21] Most record unusual operations such as ovariotomies (removing the ovaries), hysterectomies (removing the womb), limb amputations, and brain surgery.

The films were shot in a room in Doyen's clinic that was specifically furnished for the purpose. The walls, for instance, were covered with special paint to prevent unintended light reflection, and, in addition to the room's natural light source, four electric lamps assured sufficient lighting for both surgeons and cameras. The setting in all of Doyen's films is simple: The surgeon and his assistants stand next to the operating table, facing the camera (French film historian Thierry Lefèbvre has referred to this setting as a fixed choreography of *opérateurs,* the French word designating both "surgeon" and "camera persons"[22]). Two cameras record the actions of the surgeon, whose face is hardly ever seen. Patients appear as anonymous, faceless objects—only body parts essential for surgery are shown. This particular setting underscores the purpose of the films; all attention is geared toward recording the medical act, while the roles of both patients and surgeons are downplayed. Doyen held strict views about the editing of his operation films. Because their purpose was solely didactic, he insisted on inserting still shots of written explanations preceding the moving images of actual cutting. Although Doyen was an experienced surgeon, he claimed his films helped further refine his manual dexterity because they reminded him of specific actions or gestures he had forgotten in the tension of the operation. It was Doyen's strong conviction that public screenings of his films should always be accompanied by commentary from an experienced surgeon who had been present at the operation; without such explanation, he argued, the films would fail to be instructive. At national and international conferences, Doyen himself provided commentary for his films. Yet Doyen's fame extended beyond medical circles; at the 1900 World Expo in Paris, his films were

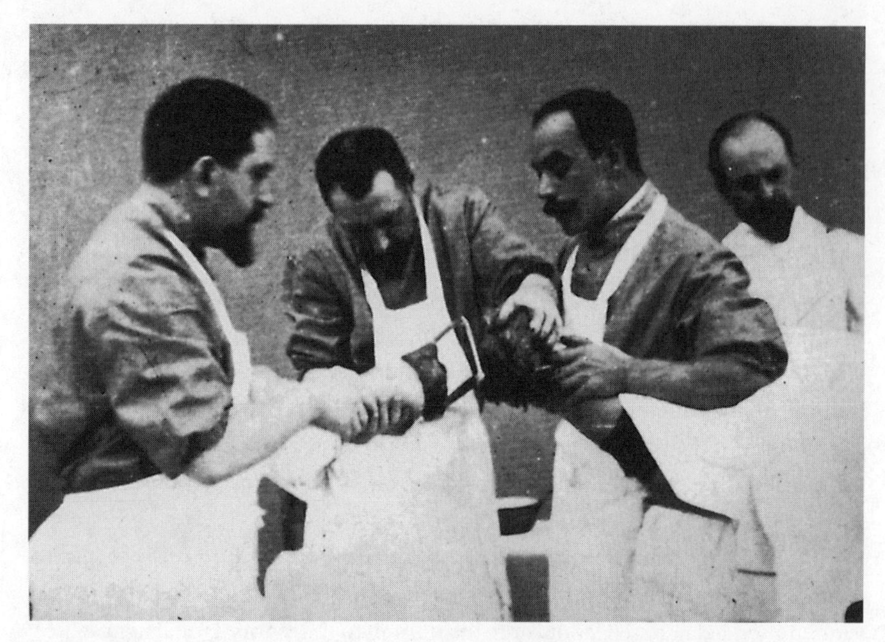

Fig. 3. Doctor Doyen and his colleagues amputating a leg. Thierry Lefèbvre, private collection.

a popular attraction. Doyen was also invited to organize special screenings for Monaco's royal family and their guests.

Undoubtedly, the most remarkable and well-known film in Doyen's collection is the one featuring the separation of the conjoined twins Doodica and Radica Neik, of which only a short fragment of the original eight minutes shot in 1902 has survived.[23] The Neik sisters, born in India, were attached through the skin of their bellies. They were one of the main attractions of Barnum and Bailey's circus, which traveled across Europe at that time. When, at age twelve, Doodica became infected with tuberculosis, surgical separation seemed inevitable in order to prevent her sister from becoming infected as well. This operation took place in Dr. Doyen's clinic and was recorded on film. An image of the two girls peacefully sleeping next to each other in the same bed, as separate persons, is one of the few shots remaining. Although the operation was considered a success, Doodica died a week afterwards; her sister would live for another year before dying (of tuberculosis). Only a few weeks after the operation, the film was screened at conferences in Paris and Berlin. Doyen also published an extensive description of the operation in a medical journal, most likely a basic outline of his oral commentary at the Paris and Berlin screenings.[24]

Doyen regarded his film as a record of a unique medical event, but his opinion was not shared by everyone. One of his colleagues, Dr. Legrain, surgical chief at the Ville-Évrard hospital, publicly accused Doyen of charlatanism and clowning.[25] Legrain felt that the film was of dubious character and harmful to the profession. Doyen responded furiously until he found out what had triggered his colleague's dismay. Without Doyen's consent, one of his opérateurs, Parnaland, had sold copies of the film about the Neik sisters to an impresario for 600 francs and, as it turned out, it was not the only operation film he had sold. These illegal copies were shown in various places, such as coffee houses and at fairs, to a lay audience, although Doyen strongly felt they were only appropriate for a medically trained audience. Doyen started a court battle against his cameraman and, at a time when the issue of film copyrights was still fuzzy, against the film distributor, Pathé. The court battle lasted for years. It took until 1905 for Parnaland and Pathé to be legally required to return the copies and forced to pay substantial compensation.[26] Doyen subsequently offered the exclusive distribution rights to his films to the Société Géneral des Cinemathographes Éclipse on the condition that they would only be shown for medical and educational purposes.[27] In practice, the court ruling meant that his films could only be screened at medical schools, but, as the society's lending records show, medical schools were not interested in Doyen's collection.

Doyen's movies mainly found an audience outside the medical domain. No doubt his decision to film spectacular operations using advanced camera techniques contributed to the unintended popularity of the films. Ironically, the film of the Neik sisters' surgical separation more or less catered to the same entertainment niche in which the sisters had formerly functioned as a live attraction, the film version replacing the life *monstrum*. The surgical dexterity depicted in Doyen's films elicited both amazement and disgust among lay viewers. In short, the films served every purpose except instruction. Eugène Doyen's efforts to use the new medium for didactic purposes failed miserably; much to his chagrin, his movies were viewed as entertainment. His influence was confined to the walls of the operating room; outside this domain, the opérateurs and distributors took charge.

1954: THE FRISIAN CONJOINED TWINS

Like most European countries, the Netherlands has a long-standing tradition of capturing medical operations on celluloid. The Dutch Film Museum owns a small collection of operation films produced between 1910 and 1940, mostly single-camera films of complex surgical interventions.[28] Although these films, without exception, were made to introduce future doctors to the professional skills of experienced surgeons, the educational value of the films proved rather limited. Due to bad lighting, the limited maneuverability of early film cameras, and the medically required distance between the fixed camera and the surgeon's activities, the details of the surgeon's work are poorly captured. After the Second World War, the 16 mm camera began to be deployed on a regular basis, and as film technology advanced, the pictures improved substantially. The introduction of color film in the 1950s, for instance, did much to sharpen the imagery and heighten its realistic character. By utilizing an additional hand-held camera, it became possible to alternate between close-up shots of the operation and shots of the operation theater, which added narrative depth. Improved editing techniques; the insertion of diagrams, photos, and X rays; and the use of voice-over significantly enhanced the narrative flow, signaling the coming-of-age of the operation film.

This evolution toward a more narrative operation documentary, made possible by increasingly complex film techniques, was closely intertwined with changes related to the production and targeted audience of operation films. If operation films in the early decades of the twentieth century were commonly the initiative of individual doctors or specialists who happened to be interested in the medium, their production was gradually taken over by professional organizations. In 1950, Dutch hospitals and universities began to commission University Film (UNFI), a nonprofit organization, to produce professional scientific documentaries. Initially, the audiences for these films consisted exclusively of experts and scientists. The Dutch chapter of the International Scientific Film Association (ISFA) had been founded in 1949, and their medical section became one of the most active groups in the association.[29] Periodically, they organized film screenings for their members, and almost half the films shown involved operations. As the 1950s progressed, these screenings became increasingly popular; initially closed to nonmembers, pressure mounted to open up their screenings—first to all medical professionals, and later also to their guests. A growing lay interest in operation films is clearly reflected in their production and format as more cinematographic means were deployed to produce a balanced and rounded case story. Although operation films of the 1950s still catered primarily to an audience of medical professionals, they tended to appeal to a wider audience as well.

The surgical separation of conjoined twins remained a favorite subject throughout the twentieth-century history of the operation film, both in Europe and the United States. A revealing example from 1954 is *The Frisian Conjoined Twins,* a Dutch production that was distributed on both sides of the Atlantic.[30] No longer a bare documentation of the medical act, this film relies on a variety of narrative techniques to tell a full-fledged story of a case history. The documentary starts by showing drawings of the complicated birth of the sisters Tsjitske and Folkje De Vries, whose bellies are connected by a large ligament. From drawings and diagrams, the viewer learns that the girls share one placenta, and hence one navel, but that they each have their own vital organs, such as heart, liver, bladder, and intestines. A snapshot of the twins shows how they looked at birth. At five months, they are ready to be operated on. Before the actual operation begins, a voice-over explains its logistics, and this is followed by the visual results of all relevant diagnostic tests: an ECG shows the synchronous heartbeat of the two children, a blood-circulation test demonstrates Folkje to have an abnormal blood-sugar level, and an X ray using a contrast medium reveals that their livers, though linked by some sort of bridge, still function as separate organs. A few shots of the naked, pink babies, held against a black background, illustrate how all of their movements require mutual coordination. Following a diagram of the layout of the operating room, the first incisions mark the beginning of the operation. Twenty minutes into the film, the viewer, like the surgeon, is well prepared for the surgical act.

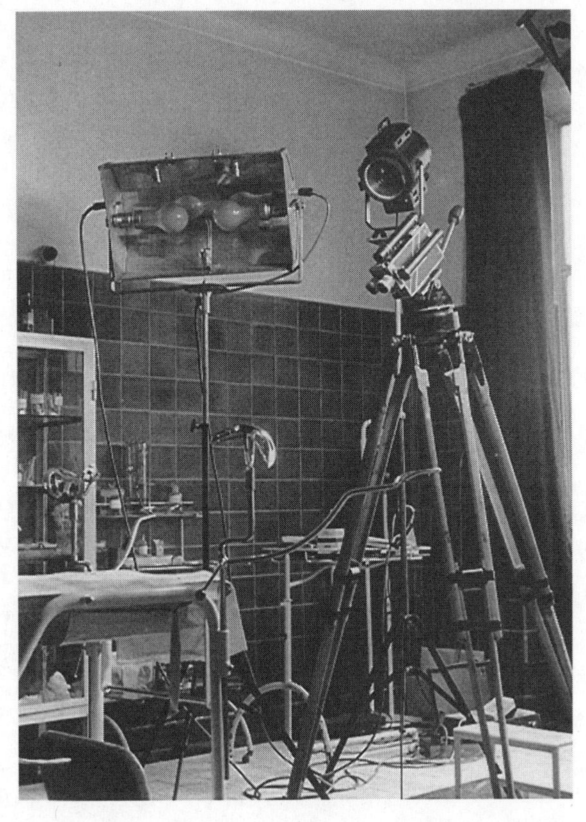

Fig. 4. Operation room with film equipment, 1950s. Archive for Film and Science, Netherlands. Courtesy of Bert Hogenkamp.

The narrative character of the film is enhanced by the use of voice-over and visual indications of the progression of time. At regular intervals, shots of a ticking clock structure the operation and help build tension. The distance between the camera and the operating table, and the many busy hands in the way prevent close scrutiny of the procedure's technical details, yet the alternating close-ups and general shots provide a good overall impression of how the operation is progressing. After the actual separation, the two babies are put on separate tables, their intestines are put back in place, and finally their wounds are sutured. A voice-over serves partly to elucidate technical details and partly to add a light touch to the story. Because the twins had shared one navel, Folkje needs to have one custom-made; the commentator remarks, "A belly without a navel is like a desert without an oasis." A group portrait of the medical staff signals the completion of the operation proper. The documentary, however, goes on to address the post-clinical phase as well: against a black background, we see Folkje and Tsjitske performing all kinds of movements to test their sensorial systems, shots that are meant to convince the viewer of the operation's successful outcome. The film concludes with a shot of two smiling nurses, each cautiously holding up a prettily dressed baby. "All participants in this operation wish both the children and their parents good luck," the voice-over reads, as the credits signal the end of the documentary.

As my account suggests, surgical intervention is hardly the main subject of the film; what matters is the full story, from birth and diagnosis to operation and aftercare. *The Frisian Conjoined Twins* is not simply a record of an operation, but a carefully edited film that documents a specific medical case from beginning to end.[31] The operation film, in other words, has become a film operation, geared toward telling a fully developed linear story illustrated by a host of verbal and visual data: drawings, diagrams, photos, test results, voice-overs, and operation shots from various angles. Patients are no longer anonymous body parts but have faces and names. The crew in charge of the medical operation is showcased in a group portrait after the successful procedure. As much as it is a medical document, this film is also a promotional vehicle for the medical profession—the Guild of Surgeons, in particular. Significantly, as the credits reveal, the film also has a commercial sponsor: the Frisian Dairy Cooperative, which sells milk and baby food. The need for corporate sponsorship suggests that the production of a documentary made with advanced film techniques is a costly affair, especially when its distribution is largely restricted to a nonprofit circuit.

The documentary on the Frisian twins is indicative of a general shift in the development of the operation film; it suggests that purely scientific interest in the visual recording of surgical activity gave way to a hybrid interest in its documentation, information, dramatization, and promotion. As in the case of the films of Dr. Doyen in the early part of the century, the didactical value of *The Frisian Conjoined Twins* is disputable. Various aspects—such as the narrative structure, the use of editing and animation techniques, and the representation of doctors and nurses as characters—contribute to the dramatization of the operation film,

Fig. 5. The Frisian conjoined twins, after separation, 1954. Archive for Film and Science, Netherlands. Courtesy of Bert Hogenkamp.

suggesting an active effort to open it up to a larger audience. While this trend was explicitly opposed by some ISFA members, especially those who prioritized the scientific content of these films, surgeons were keenly aware of the film's promotional significance for advancing their professional status. The debate unfolding in the 1950s is partially recorded in the minutes and reports of the Dutch chapter of the ISFA.[32] Doctors and specialists wondered whether it was appropriate to mix medical record with drama and entertainment, and whether a medical-scientific film should be manipulated to fit the requirements of a dramatic format. Because laypersons were increasingly allowed to visit screenings, questions arose, such as: Should narrative techniques outweigh operation techniques? Is sponsoring or commercial funding permitted? Issues that had arisen as early as Dr. Doyen's time became urgent again in the 1950s; they became even more pressing after the operation-film genre was incorporated into television—a medium geared, first and foremost, toward entertaining a mass audience.

1995: THE TELEVISED OPERATION OF DAO AND DUAN

Since the early 1990s, the medical documentary has gained substantial popularity as a television genre, both in Western Europe and in North America. It took quite a while for television producers to discover the popular appeal of medicine in general, and of surgery in particular. After all, operations have many of the essential ingredients that sustain drama: shrewd heroes, helpless victims, state-of-the-art technology, and a struggle for life and death. In the operating theater, emotion, suspense, and excitement are apparently so palpable that broadcasters have begun to produce programs featuring live and recorded operations.[33] A factor contributing to the growing popularity of operation documentaries as a television genre may be the immanent convergence of media and medical technologies. The camera and the monitor have become indispensable operating aids in modern surgery. Endoscopic techniques, commonly applied these days, allow viewers to visually track the surgeon's activities inside the human body, as the camera's eye and the surgeon's eye have fused into one and the same instrument. As we shall see in chapter 4, minute endoscopic cameras provide detailed, high-density shots of the interior body. Picture quality, transmission speed, and the availability of various types of imagery (such as MRI, endoscopic, or CT) have all improved significantly as a result of the digitization of technology. Today, media technology in operating theaters is no longer an intrusive agent but has, on the contrary, become rather ubiquitous. The convergence of media and medical technologies have resulted in an enhanced sense of the real—the suggestion of an increased transparency—when it comes to capturing the activities and anxieties involved in medical operations. Or, as professor of preventative medicine Catherine Belling has argued, "Spectacle and storytelling overlap with the material effect of surgery on real bodies."[34]

In most cases, televised operations are the coproductions of public (or commercial) broadcasting companies and hospitals (or professional medical organizations). The apparent aim of these programs is still the recording of operations, albeit no longer to educate medical specialists but to inform the general public. If, as we have seen, the operation films of the 1950s sought to represent a medical case comprehensively, thus catering to a wider audience, from the 1990s onwards, documentary producers have definitely aimed at extending their narrative and dramatic appeal. The subject of most contemporary televised operations is not strictly the medical procedure, nor is it the attempt to depict a medical case history. Paramount to the success of these programs is their human-interest angle: what matters most is representing patients and doctors as humans. Names become faces and faces become individuals who have relatives and personal histories. A surgeon is no longer purely an expert professional, but a human being with feelings, views, and empathy. Televised operations, then, are an excellent framework for promoting the skills, interests, and advancements of the medical profession, and the eagerness of hospitals and professional organizations to cooperate with television stations underscores the promotional relevance of these programs for elevating the status and prestige of their field. The interest of surgeons in attracting attention to their work and the interest of broadcasting organizations in seeking the largest audience possible seem to go hand in hand.

The 1995 American documentary *Siamese Twins,* produced for and broadcast by PBS, offers an eminent example of how media and medical interests have merged.[35] Although the program is presented as a medical documentary, it surpasses the limitations of that genre; as the opening statement suggests, viewers will be witnessing "a story of love and courage." The two-year-old Thai sisters, Dao and Duan (no surname is given), are first shown at the Philadelphia airport, where their prospective adoptive parents, Barbara and David Headley, are eagerly awaiting their arrival. The girls' entire lower bodies are conjoined from the mid-torso down, and they share three legs.[36] A voice-over frames their story as a conventional immigrant saga: "They begin a new life in a strange land. Their future is uncertain and full of risk." The twins' surgical separation is supposed to be the program's main subject, yet only a quarter is devoted to the actual operation. It is not the conjoined twins but the surgeons and advanced medical technology who play the leading parts in this drama.

Two lengthy operations are necessary to separate Dao and Duan, and these two procedures provide the basic narrative frame for the program. The first operation is needed to split their one spinal cord into two and to insert inflated balloons under the skin of their legs in order to artificially stretch the skin. The second operation—three months of stress and suffering later—will actually separate the bodies; Dao will get the third leg and the largest part of their shared bladder. Before and during the operations, suspense is heightened through comments like "this is a point of no return" and "today is the most important day in the twins' lives . . . one or both could die." Shots of a ticking clock visualize the progression of fourteen tense hours in the operating theater. A camera right above the operating table gives viewers more or less the surgeon's perspective, registering the intervention seemingly through his eyes. Yet, frontal shots of the opened-up bodies are used sparsely and are quickly interlaced with bird's-eye shots of the operating theater, which is populated by some thirty busy specialists, assistants, and technicians. While scrubbing before the operation, individual surgeons explain their specific contributions, and a few animations clarify the technical details. As soon as the operation is over, we can see and hear the surgeons casually discussing the result; they are happy that "the children are now able to live their lives as separate persons."

In this documentary, technology plays at least as heroic a part as do the surgeons. Dao and Duan's predicament certainly poses a challenge, but, as Dr. O'Neill, one of the physicians interviewed, claims, "they are fortunate to be born in the age of high-tech medicine." Without sophisticated techniques like MRI and CT scanning, this operation would have been impossible. "Modern imaging techniques will allow the team to visualize the complex anatomy before contemplating surgery," the voice-over explains. Perhaps inadvertently, these new techniques doubly expose the twin's private interior bodies. The camera not only depicts the girls' insides through direct close-ups of the operating table, but also indirectly, via scans. Referring to specific MRI and CT images, surgeons point out the organs the girls share, after which they use a pencil to draw their exact locations on the girls' actual skin. The diagnostic scans seem to legitimize the spectacle as a scientifically instructive lesson, yet at the same time they degrade the girls, whose flesh becomes no more than a superficial layer hiding their actual organs. Even though the camera records them in their naked vulnerability, viewers have no reason to be ashamed of their voyeurism; after all, the medical scans legitimate the spectacle.

It is obvious that *Siamese Twins* promotes both medical expertise and medical technology, but it may be less clear how the program directly benefits any of the parties involved. Needless to say, an operation like Dao and Duan's does not come cheap, and American hospitals are anything but charitable institutions; financial matters, however, are never explicitly addressed in this documentary. The Children's Hospital of Philadelphia, the voice-over explains, will not charge for the use of the imaging technology, while the surgeons generously donate their expertise. In exchange for their work, the surgeons have the opportunity to "sell" their expertise through the television program. Indirectly, though, the program aims to generate funds for the operation itself. At the end of the documentary, the adoptive mother of the twins explains that their insurance company will cover neither the cost of these operations nor the expenses of operations that still lie ahead. Duan, who received the third leg, will have to undergo an amputation of her right leg that, as it turns out, failed to grow, and Dao needs lifelong medication in addition to more surgery, prostheses, and extensive physical therapy. In order to recoup some of the gigantic costs, the Headleys have established a fund, and viewers are implicitly encouraged to send in donations. The ending of the documentary reveals the bitter irony of the new situation the girls find themselves in. The separation has wiped out their prime exchange value: being conjoined twins. After the operation, Dao and Duan are simply two seriously disabled Asian American children. The operation has "normalized" their handicap, and the media generally pay little attention to children with ordinary handicaps.

At first sight, the documentary may appear to be an innocent, if blatant, advertisement for doctors and their technology; below a thin layer of human empathy, the ideological undercurrents are hardly disguised. The dominant Western ideals of superior technology and socio-medical justice are sharply pitted against the technological and social backwardness of the non-Western world. In more than one way, the story of Dao and Duan echoes the story of Eng and Chang Bunker, the "original" Siamese twins—a resemblance that the producers are keen to exploit.

The voice-over presents the story of Chang-Eng as a case of the American Dream come true. The immigrant brothers, after all, were able to build prosperous lives in the United States, their only misfortune being that the technology required to separate them was not yet available. Today, their separation would have been a fairly simple surgical procedure, as Dr. O'Neill explains. Implicit in his conclusion is the suggestion that anomalies like conjoined twins are practically nonexistent in the contemporary Western world on account of its advanced technology; early detection by medical imaging techniques and diagnostic tests virtually prevents these creatures from being born.[37] Conjoined or "Siamese" twins are thus a doubly exotic phenomenon, which occurs only in the world of the ethnic Other, while it can be treated only in high-tech Western countries.

The most striking correspondence between the Bunker brothers and the sisters Dao and Duan remains unstated in the documentary. Chang-Eng had to exhibit their bodies at fairs and traveling circuses, first to make money for their contractors, and later to support their own families, while Dao and Duan had to exhibit the operation that turned them into separate beings in order to pay for it. Chang-Eng were able to redeem themselves from the managers and showmen who made a fortune from their public performances; Dao and Duan (who were too young to give consent) pay for their "liberation" by publicly displaying their suffering. Whereas the role of doctors in the nineteenth-century freak show was to authenticate medical curiosities, in this new cultural configuration, doctors have taken center stage and become part and parcel of the freak show. In response to the documentary, medical ethicists David Clark and Catherine Myser rightly raise the question, "Does the virtualized universe of television make what was once an obviously exploitative spectacle into something safe, that is, as hygienic and enlightened as the theater of surgery?"[38] *Siamese Twins* is not simply a return to the nineteenth-century freak show in which the Bunkers took part; it is a grotesque intensification of it. By and large, the freak show functioned as a straightforward pay-per-view spectacle of the aberrations of nature. The 1995 documentary unashamedly invites millions of viewers to gaze at human deformity disguised as medical information and human interest.

The late-twentieth-century televised-operation genre relies on a mixture of medical technology, media technology, and commercial and cultural ideology, the separate elements of which are hard to identify. In 1902, Dr. Doyen was furious when he found out his films were being shown at fairs, yet in his film of the separation of the Neik sisters, the patients are hardly recognizable and the surgeons figure as anonymous pairs of hands. The 1954 production *The Frisian Conjoined Twins* turned the sisters De Vries into identifiable medical subjects and devoted modest attention to the medical team. It relied more heavily on narrative film techniques, while mixing educational, promotional, and commercial interests. At the end of the twentieth century, operation films on conjoined twins have evolved into full-blown television dramas about doctors, technology, and the blessings of Western culture, while the twins—as medical subjects—have been relegated to the sideline. In some of its dramatic content, the documentary *Siamese Twins* is almost indistinguishable from actual drama. A careful use of edited images, real dialogue, and narrative techniques turn the deformed body and the dangerous intervention into an alluring and exciting spectacle, in which, ostensibly, no other goal than the childrens' best interest is served. The medicalization of the conjoined twins, it seems, is now complete, and appears to override every appearance of voyeurism or exploitation. Foucault would term this the "normalization" of the medical gaze, as it is shared by millions of viewers and thus renders these public spectacles "normal."

While examples of blatantly foregrounded medical and technological interests are fairly easy to pinpoint in *Siamese Twins,* the role of the media is much harder to identify. As Guy Debord has argued, it is precisely the invisibility of the camera, and of televised production in general, that is characteristic of late-twentieth-century media spectacle. Put plainly, the aim of this documentary is to enable its actual broadcast—without television, no publicity, and without publicity, no money for further operations. The mediated event, as Debord suggests, has replaced the real event. Since the nineteenth-century freak show was first incorporated by medical discourse, the media have integrated this phenomenon into our culture at large. Television is the road show or circus of the late twentieth century, and media producers have taken the place of freak-show managers. Naturally, their motives are noble and their compassion is sincere. The point of my analysis, however, is not to reveal individual motives or intentions but to explain how the format and content of a popular genre both reflect and construct specific norms and values concerning "deviant bodies" in our everyday culture. Televised freak shows are a legitimate, if not legitimizing, part of our advanced civilized modern society.

2001: A MEDICAL-MEDIA-ETHICAL DILEMMA

A recent British documentary shows that there are alternatives to the technique and style used in *Siamese Twins.* In September 2000, a court case in Manchester raised an individual medical-ethical dilemma to the status of national ethical debate. The parents of newborn conjoined twins Mary and Jody (pseudonyms) filed suit against the hospital that wanted to surgically separate the twins. Jody fully depended on the blood-circulation system of her sister; left untreated, the situation would inevitably result in the death of both twins. Separating Jody from Mary could at least spare the life of one of the children; the parents' religiously motivated stance, however, was that saving one child could never justify the killing of the other. The judge concurred with the hospital's medical judgment, and allowed the operation to proceed—an operation in which Jody died, as expected. The court case highlighted important medical-ethical questions: Is there a medical imperative to separate conjoined twins? Can otherwise healthy conjoined twins live a normal, happy life without being separated? Who has the right to decide on the future of newborn conjoined babies—parents or doctors?

In the midst of the heated public debate following the court case, the BBC broadcast *Conjoined Twins,* a documentary that sensitively visualized the complexity of Mary and Jody's predicament.[39] There are various narratives at the heart of this documentary. First, we witness the hospitalization and successful surgical separation of conjoined twins in Cape Town, South Africa. At the request of their parents, doctors had successfully operated on the babies, who leave the hospital as two individuals. Yet this is not the only story. We also hear the tragic tale of the Irish twins Katie and Eilish, who both lose their lives during a surgical attempt at separation. Next, we see shots of the eleven-year-old Californian sisters Abigail and Brittany. Conjoined from the shoulders down, they live the happy and carefree lives of preteens: we see them swimming, horseback riding, and jokingly discussing boys. The story of the Russian sisters Masja and Dasja illustrates the opposite outcome: even after fifty years of forced physical companionship, they are still frowned upon by neighborhood residents and relatives. Finally, we are introduced to the adult twins Lori and Reba from Pennsylvania, who were born with their heads joined—an inoperable medical condition. Their handicap, however, does not prevent Lori from posing as a model for magazines and art journals. The most remarkable feature of this BBC documentary is its refusal to take a stance for or against the medical imperative; contextualizing and intertwining various cases, the producers carefully balance medical solutions with other options, each time highlighting different factors contributing to the welfare of the twins.

Conjoined Twins not only highlights an intricate medical-ethical dilemma but addresses the question of media ethics as well. Whereas, in the case of Mary and Jody, the judge weighed the pros and cons of a medical injunction, the television producers question the imperative of surgical intervention by alternating the stories of happy and unhappy conjoined twins. Rather than foregrounding the high-tech medical spectacle, they use the medium to offer a nuanced view of malformed bodies and persons living with a physical handicap. The viewer's gaze is diverted from the authoritative eye of medicine to include nonmedical perspectives. Surgeons and medical ethicists comment on the technical and moral considerations involved in making such life-or-death decisions, but more important than the voices of professionals are the voices of the actual twins. In this documentary, conjoined twins are not just talked *about* or publicly displayed—they are autonomous subjects who speak for themselves.

The operation film, in the past century, has largely manifested itself as a mediated medical spectacle; the professional interests of medicine and media have distinctly dominated the public image of conjoined twins as a medical curiosity. As we can conclude from the BBC example, documentary makers also have a responsibility in the medical-ethical dilemma. The way in which conjoined twins are framed and depicted defines how they are looked upon and treated by the audience—either as helpless surgical objects or as autonomous human subjects. Doctors, of course, have a professional responsibility: deciding what is best in such precarious situations. If they allow camera crews into the hospital, they may yield power over the representation of their work to the media. Conversely, when the media enter the domain of the surgeon, they may uncritically submit themselves to medical authority. The coalescence of two professional groups with diverse interests may result in a mutually profitable coalition of forces, such as illustrated by the American documentary *Siamese Twins,* or it may lead to a confrontation of professional intentions, as evidenced by the example of Dr. Doyen.

Naturally, in the construction of public images, power is not distributed equally. But when the representation of handicapped people is at stake, producers and doctors become jointly responsible for the representation of their subject. In the summer of 2003, when Iranian twins Ladan and Laleh, conjoined at the head, were surgically separated in Singapore, newspapers all over the world held their audience in breathless expectation. "The world mourns," they reported when the unfortunate sisters did not survive the complex surgical procedure. Several doctors openly questioned whether all this publicity had not actually harmed the twins by putting too much stress on the doctors, or by raising expectations to an unrealistic level. Every time a new medical case is taken into the limelight, a medical-ethical dilemma becomes a media-ethical dilemma. Although in the public domain professional responsibilities often appear inseparably intertwined, it is vitally important to disentangle the shared interests of doctors and media producers in order to anticipate the real consequences for the subjects involved.

CHAPTER 3

BODYWORLDS: THE ART OF PLASTINATED CADAVERS

IN THE 1950S, WHEN SYNTHETIC MATERIALS HAD RECENTLY BEEN INTRO-DUCED, PEOPLE USED TO ADMIRE PLASTIC TULIPS FOR THEIR REALISTIC QUALITY. CONSUMERS WERE CHARMED BY THE OBVIOUS ADVANTAGES OF THESE FAKE FLOWERS: THEY NEVER WITHERED, AND EVERY TULIP LOOKED ABSOLUTELY PERFECT. WHEN I BUY A BOUQUET OF REAL TULIPS THESE DAYS, IT STRIKES ME HOW MUCH THEY RESEMBLE PLASTIC ONES. BY AND LARGE, THE FAMOUS DUTCH TULIP IS NO LONGER AN EXCLUSIVE PRODUCT OF NATURE, but depends increasingly on treatment with chemical and biotechnological means for its cultivation. Again, the advantages are obvious: the flowers remain fresh much longer, and every single tulip meets the requirements of standardized size, shape, and color. Whereas before, we wanted the artificial object to look like a real one, we have now entered an era in which we want the real object to look like "perfect nature." We are no longer satisfied with a plastic imitation of an organic object, yet neither are we satisfied with nature's own imperfect products. So we tinker with flowers until they meet our aesthetic standards. The contemporary tulip, in other words, has become an amalgam of organic material, cultural norms, and technological tooling.

This new preference for the enhancement—instead of imitation—of nature also pertains to the human body. Dentists who, in the 1960s, did not think twice about pulling a patient's teeth and replacing them with a set of dentures (cheap and low-maintenance) now make every effort to save the original ivories. They have extensive collections of tools and materials at their disposal to perfect our pearly whites, until they resemble the (retouched) teeth of fashion models in magazine photos. Our physical appearance can be optimized by plastic surgery, anabolic steroids, and, perhaps in the near future, genetic therapy. "Natural silicone breasts" is no longer an oxymoron but an indication of a reality in which female bodies are reshaped by cultural norms with the help of advanced technology. The desire for a manipulable body perfectly fits a material, technological culture in which imitation has been replaced by modification. Like the tulip, the body has become a mixture of nature and artifice.

If the living body has become a mix of nature and artifice, it is no great surprise to find that the dead body has too. In the past twenty years, Gunther von Hagens, a German anatomist from Heidelberg, has developed a preservation technique that he has dubbed "plastination." It involves a sequence of specific chemical treatment of the corpse, which is then modeled into a sculpture by the anatomist's hand and scalpel. The resulting anatomical object looks like a conflation of opened-up mummy, skinned corpse, and artistic sculpture. Von Hagens calls his collection of cadavers "anatomical art," which he defines as "the aesthetic and instructive representation of the inside of the body."[1] After its first public showing, in Japan, von Hagens's remarkable collection Körperwelten (Bodyworlds) was exhibited in Mannheim, Germany (1997–98), Vienna (1999), Brussels (2001), and London (2002). The first exhibition in this series, in Mannheim, lasted four months and attracted almost one million visitors—an exorbitant number for what was advertised as a scientific exhibition—and the Vienna event was kept open twenty-four hours a day, seven days a week, to accommodate all visitors. Even exhibits in major art museums devoted to the work of canonized painters seldom receive this much popular attention.

What, then, makes the plastinated bodies so fascinating? Why did Bodyworlds become such a success? Evidently, in our increasingly medicalized society, people's interest in the human body has risen in proportion to their interest in its normally hidden dimension. And yet, other anatomical-pathological museums in Europe have offered glimpses inside the body without attracting anywhere near the number of visitors Bodyworlds has. One factor contributing to its popularity may well have been the public debate, fanned by the German media, about the ethicality of this exhibition.[2] Newspapers and television shows raised the question of whether the display of real human cadavers was indeed legitimate, and if so, for what purposes? Did Bodyworlds serve any scientific goal at all, or was its prime intention the display of artistic objects? Undoubtedly, this media attention drew more visitors to Mannheim, but it does not fully account for the exhibition's immense popularity.

The appeal of Bodyworlds and the controversy surrounding it can be properly understood only if we approach the phenomenon from a historical perspective. Von Hagens's plastinated cadavers fit the long-standing scientific tradition of anatomical body production, as well as the artistic conventions of anatomical representation. By the same token, the remarkable exhibition setting can be retraced to its cultural roots in public anatomy lessons and the artful display of body parts in glass bottles, as shown in anatomical museums. From the history of anatomy, we have learned that anatomical practices, objects, and representations have always been an intricate mixture of science and art, and a hybrid of medical instruction and popular entertainment. During the Mannheim exhibition, the ethical debate centered primarily on the question of whether plastination should be looked upon as science or art, instruction or entertainment. What makes von Hagens's anatomical art controversial, though, is not that it cannot be clearly classified, but that it defies the very categories in which ethical judgments are grounded.

THE HISTORICAL TRADITION OF ANATOMICAL BODIES AND MODELS

Throughout its history, anatomical practice has tried to reconcile two contradictory requirements of medical education: authenticity and didactic value.[3] On the one hand, the anatomical body should consist of real flesh, so that cutting into a cadaver allows future doctors to experience the complexity of a living human body.

Yet working with dead bodies has a distinct drawback: it is difficult to demonstrate certain aspects of physiology, such as blood circulation or the complex web of muscular tissue. In order for students to conceptualize anatomical structures, an anatomical body should be pliable so that particular aspects can be singled out. Body models, shaped and sculpted to reveal distinct parts or features, have served as teaching aids in medical schools since the early modern periods. The advantage of models is that certain physiological features can be disproportionately accentuated in order to convey particular anatomical insights. An obvious disadvantage of body models is that they do not give students a feel for organic texture. From the time of Andreas Vesalius to the days of von Hagens, we see anatomists struggle to combine a preference for authentic bodies with the educational advantages of body models.[4]

The need to preserve corpses for more than several days, and anatomists' desire to demonstrate particular physiological features stimulated the invention of better preservation methods. Between the early twelfth and sixteenth centuries, various techniques for embalming or preserving corpses were experimented with.[5] The Dutch anatomist Frederick Ruysch (1658–1731), successor to the illustrious Dr. Tulp, developed unprecedented standards for the preservation and display of bodies. He injected the veins with a mixture of talc, tallow, cinnabar, oil of lavender, and colored pigments, the precise recipe of which he kept secret. As a result, the body would last much longer, sometimes up to a full year, and dissection was less messy, due to the replacement of blood by preservative. Yet Ruysch's technique did more than ameliorate the material preconditions for dissection. His technique allowed for a new kind of anatomical artifact—a work of art rather than a scientific work object. As art historian Julie V. Hansen observes, Ruysch created "a new aesthetic of anatomy that melded the acts of demonstration and display with the stylistic and emblematic meaning of Vanitas art."[6] Besides performing public dissections, Frederick Ruysch built up a collection of body parts, such as hands, limbs, and heads, carefully conserving each in a separate glass jar. To enliven his objects and disguise the brutality of death and dismemberment, he embellished the compartmentalized cadavers with flowers or garments. His favorite displays were the little bodies of fetuses or stillborn babies that he clothed with scarves and embroidered baby hats, replacing their eyes with glass to make them look like innocent infants. Even though Ruysch was one of the most respected Dutch anatomists of the seventeenth century, he is consistently referred to today as an artist who elevated anatomical bodies to the status of sculpture and painting. As Hansen argues, "Under Ruysch's hand, the body was not dead, it was nature revealed, admired as the handiwork of God, the invisible made visible."[7]

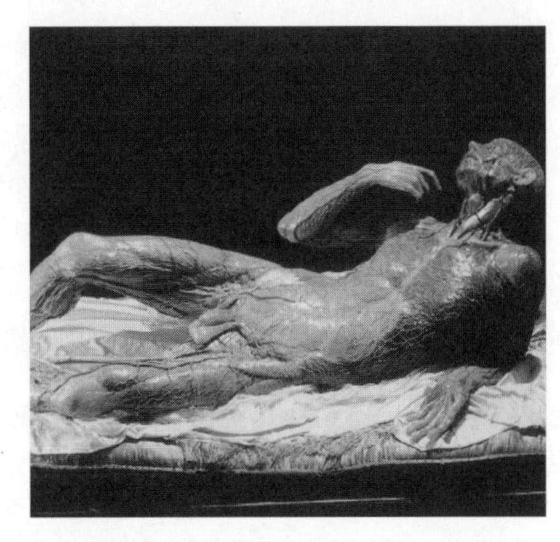

Fig. 6. Wax model "Lo Spellato," La Specola, Florence, Italy.

British cultural historian Ludmilla Jordanova emphasizes that the conflicting requirements of authenticity and didactic value emerge repeatedly throughout the history of anatomical artifacts.[8] After the Renaissance, medical education increasingly called for hands-on practice. Greater demand and tougher laws on obtaining cadavers forced anatomists to look for body substitutes.[9] As practical solutions to the shortage of real cadavers led to the creation of "fake bodies" in the seventeenth and eighteenth century, these models were subjected to equal standards of accuracy, durability, and technical flexibility. The development of wax models catered to these educational needs, and had some advantages over real bodies.[10] Beeswax has the unique quality of resembling organic texture, while it is also fully pliable. Parts of a wax model could be removed to allow the student to view the organic complexity, or to manipulate individual parts and organs. In the second half of the eighteenth century, Bolognese sculptors like Ercole Lelli and Giorgio Morandi and Florentine masters like Leopoldo Marc Aurelio Caldani, Felice Fontana, and Giovanni Battista Piranesi raised the craft of wax modeling to an art; some began to be commissioned by royal Maecenases and their models were bought up by private collectors.[11] From clinical-instructional settings, the wax models passed to private collections and later to museums, where they can still be admired.[12] After wax, several other materials were used for the production of models.[13]

The invention of new chemical techniques, particularly the application of formaldehyde in the nineteenth century, allowed anatomists to extend the preservation of cadavers and enabled students to participate in actual dissections. Dissections were no longer public events, as they had been in the Renaissance, but took place behind the closed doors of the hospital lab. Through various modes of public display—most notably formaldehyde-drenched body parts in glass bottles—we can further trace the hybrid requirements of authenticity and pedagogical value. In contrast to the embellished body parts from collections such as Ruysch's, nineteenth-century exhibitions of organs in glass jars show a preference for unadorned anatomical parts. Reproductive organs affected by sexually transmitted diseases or livers degenerated by alcoholism clearly served a double pedagogical mission. These specimens instructed doctors about the regularities and irregularities of human anatomy, yet their broader aim was to teach ordinary men and women the laws of moral behavior. The primary appeal of such anatomical collections was their focus on the aberrant—especially the monstrous aspects of pathological cases, such as embryos with spina bifida and fetuses with hydrocephalus. Although pathological creatures and "monsters" as objects of spectacle had historical roots in European fairs and curiosity cabinets from the Middle Ages, their display in anatomical exhibitions rendered them part of an authoritative medical culture.[14] Monstrosities and deformed fetuses preserved in formaldehyde commanded respect not only for the relentless power of nature (and, in its wake, the arm of God) but also for medical science, which was capable of dethroning this power. A mixture of authenticity and educational value, of titillation and moralism, characterizes the nineteenth-century specimens still to be found in anatomical museums today.

The historical tradition of anatomical modeling is pivotal to our understanding of Bodyworlds and the controversy surrounding the exhibition. The technique of plastination is both a continuation and an enhancement of a centuries-old tradition. Gunther von Hagens's assertion that the plastinates are only successful if they offer both authentic representation and educational value betrays the same tension between authenticity and the desire to instruct that has defined the manufacture of anatomical bodies and models from the early sixteenth century on.[15] Plastination, according to its inventory, manages to combine the qualities of real bodies with the advantages of body models. In his view, authenticity and didactic value—organic materiality and pedagogical plasticity—are not mutually exclusive. His plastination technique is based on a chemical treatment which renders cadavers pliable while also preventing them from decaying, and keeps the "original" body intact while still accentuating specific physiological details. By carving out the relevant parts and discarding the surrounding tissue, von Hagens highlights specific features of the body, such as muscles, bones, respiratory organs, or the heart. For instance, Bodyworlds displayed plastinated sculptures that consisted only of bone structure alongside "muscle men" that featured only muscle tissue. In other words, the plastination technique purportedly preserves the organic material, while allowing the cadaver to simultaneously function as a body model.

Put briefly, the plastination technique works as follows: The corpse is first immersed in formaldehyde, so as to stop its decay. Next, the cadaver may be prepared in one of two ways. For *Scheibeplastinate,* the body is cut into slices of sometimes less than a millimeter. These slices are chemically treated and later pressed between transparent film or glass plates. The translucent body slices almost resemble scanned images, particularly when they are displayed in lieu of the "natural" position of the respective body parts. More impressive, however, is the second type of plastinate, the *Ganzkörperplastinate.* In the production of these anatomical objects, the whole body is left intact, except that certain parts are removed, allowing others to become more visible or pronounced. After being dipped in formaldehyde, the body is submerged in a basin filled with a chemical mixture, and the body fluids are replaced with acetone. The final phase of the chemical process consists of impregnation under pressure in which the acetone is replaced by a synthetic resin.[16] The Ganzkörper are subsequently put into their final position, then treated with gas or hot air to fix the form—a form that, according to von Hagens, will last for at least two thousand years.

As opposed to wax or plastic anatomical models, the "realness" of the plastinated body is advertised as an important asset. Von Hagens stresses the authenticity of the plastinates, putting them on a higher plane than body models, which are, after all, imitations of bodies. Texts accompanying the Ganzkörper at the exhibition in Mannheim stated explicitly that the bodies were not compilations of various cadavers or partial imitations but were "real" and "intact." But what is "real"? The cadavers are manipulated with chemicals to such an extent that they can hardly be regarded as real bodies. As in the case of living bodies that have been altered by plastic surgery or anabolic steroids, it is almost impossible to use the term "authentic" in this branch of anatomy. By constantly foregrounding the realness of his cadavers, von Hagens downplays the role of chemical modification—yet it is precisely this technique that he has patented. The novelty of this method is less in the kind of chemical treatments used than in the purpose of their use: producing "natural" body models. The plastinated cadaver is thus as much an organic artifact as it is the result of technological tooling. As the engineers of genetically modified corn and wheat insist that theirs are "natural" products, von Hagens understates the process of chemical manipulation. The plastinated sculptures, however, are as much imitations of bodies as are body models, and they sometimes look less "real" (more like plastic) than do eighteenth-century wax figures.

The educational or moral function that dominated nineteenth-century anatomical collections returns explicitly in the plastinated organs and body parts. In von Hagens's plastinates, too, a claimed unmediated realism goes hand in hand with outright moralism. Bodyworlds visitors were confronted with unambiguous messages about various types of self-induced physical degeneration. Plastinates of black, tar-covered lungs were displayed alongside white, perfectly healthy ones; similarly, a healthy liver and one affected by excessive alcohol consumption were shown side by side. Bodyworlds visitors seemed particularly eager to look at the displays of physical defects, both congenital defects and those caused by disease after birth. Tumors of the liver, ulcers, enlargements of the spleen, and specimens of arteriosclerosis show the ruthless destruction of the human body. Plastinates of pathological embryonic growth, such as a fetus without a brain or a fetus with hydrocephalus, illustrated what could go wrong during the human reproductive process. Unlike the nineteenth-century bottled specimens preserved in formaldehyde, the catalogue explains, the absence of glass jars and fluids at the exhibit allows a more "authentic" or "unmediated" look at physiological reality. In this way, Bodyworlds can be understood as a direct continuation of the realist-moralist tradition in anatomical art.

At the same time, though, the Mannheim and Vienna exhibitions provided a metacommentary on the twentieth-century "nature" of the flesh. Whereas in nineteenth-century displays, natural bodies were shown to be prone to degeneration, either through God's hand or man's own immoral behavior, the plastinated cadavers celebrate the power of humankind to interfere with life and death. Von Hagens seems interested not merely in the idea that the body is an organic object that people can influence negatively by, for instance, smoking or drinking. In the course of medical history, there has been an increasing number of inventions aimed at countering—if only temporarily—physical deterioration. One of the plastinates is an explicit comment on the influence of technology on medicine. The "Orthopedic Plastinate" is covered from top to bottom with all kinds of internal and external prostheses, ranging from a metal knee and external fixtures for broken bones to a pacemaker and a replacement for a fractured jawbone. This remarkable plastinate does not only demonstrate technological progress in medical science; it entails a statement about the contemporary living body: human beings have become hybrid constructs, amalgams of organic and technological parts—cyborgs, in Donna Haraway's definition.[17] The "natural" body is no longer a given, as both longevity and quality of life can be manipulated.[18] Technological and chemical aids are promoted as "natural" extensions of the living human body, just as the process of plastination prolongs the durability of the dead body.

The history of material production teaches us that the anatomical body has always been regarded as a hybrid of art as well as science, wherein concerns for authencity and instruction tend to compete. In some periods, authencity was foregrounded; at other times, instruction. In introducing the method of plastination, von Hagens claims to have moved beyond the body-or-model dilemma, because his cadavers are both real and modifiable. He has repeatedly stressed the authencity of his anatomical creations, yet their modification is the very thing he has patented. Although in many ways his work is a continuation of age-old traditions in the material production of anatomical objects, von Hagens also adds a commentary on the "nature" of the human body. Humans are no longer subject to divine nature, as science and scientists, to a large extent, control longevity and quality of life. By the same token, a similar mixture of continued tradition and postmodern commentary can be witnessed in Bodyworlds's recasting of artistic conventions.

ANATOMICAL BODIES AS ARTISTIC REPRESENTATIONS

In one of his famous essays, art historian Erwin Panofsky argues that the rise of anatomy during the sixteenth and seventeenth centuries cannot be understood in isolation from the Renaissance in art; the history of anatomy is deeply embedded in art history.[19] He even argues that, in order to determine the scientific value of anatomical art, it should be evaluated from the perspective of the art historian. During the sixteenth century, accumulated knowledge of the body was represented visually in drawings and engravings produced by anatomists and their craftsmen. These anatomical atlases are still admired for their clear depictions of contemporary anatomical insights, but even more for their artistic qualities mirroring the conventions of early Renaissance art.[20] The illustrations in Andreas Vesalius's *De Humani Corporis Fabrica* (1543), for instance, are reminiscent of ancient Greek sculpture, with their strong bundles of muscles and round, broad-shouldered torsos.[21] A characteristic of Vesalius's engravings is that the dissected organs are surrounded by a healthy, living body, distracting from the rather repelling appearance of death; the scientific reality of the image is embellished, aestheticized, so as to make it more pleasing to the eye. Vesalius's skeletons and "muscle men" are also imprinted with the principles of sculptural tradition; although they refer to dead bodies, they pose as upright, living figures.[22] The classical conventions of Renaissance painting and sculpture determined the formative elements of his anatomical representations.[23]

Panofsky's view that artistic techniques of representation dominate and shape scientific insights is corroborated by Ludmilla Jordanova, who, in a close analysis of eighteenth-century wax models, shows how neoclassicist ideas determined the representation of scientific insights in this genre.[24] These anatomical models

Fig. 7. Wax model "Lo Spellato," La Specola, Florence, Italy.

provide perfect specimens of bodies that are partially opened up, showing, for instance, the stomach, the intestines, or the reproductive system. As in the case of Vesalius's engravings, these models are extremely vivid and their physical beauty tends to divert attention from the exposed intestines. Most of the female bodies, for example, are shown in classic Venus poses; while their main purpose is to display the reproductive organs of the female body, the wax models are expressive of the seductive goddess of love. Consequently, aesthetic standards of external appearance outshine the realistic representation of the intestines.

Historians of science Lorraine Daston and Peter Galison, focusing on nineteenth-century medical representations of the body, reframe Panofsky's argument in terms of a continuous struggle between scientific objectivity and artistic subjectivity; they historicize the concept of objectivity by what they term "mechanical" or "non-interventionist" objectivity.[25] With the arrival of new representational technologies in the nineteenth century, scientists hoped to eliminate artistic contamination. Mechanically mediated representations were thought to be conceptually distinct from earlier attempts to produce "true-to-nature" depictions of the interior body. New technologies, such as photography and later the X ray, purportedly ruled out the subjectivity of the artist, replacing it with truthful, objective imprints.[26] Yet the introduction of mechanical inscription, as Daston and Galison convincingly show, "neither created nor terminated the debate over how to depict."[27] Substituting photomechanical instruments for the engraver, they argue, did not eradicate interpretation; the photographer's very presence meant that images were mediated. New apparatuses brought the ideal of transparency closer, while promoting a new kind of objectivity—through mechanical reproduction.[28]

The anatomical artifacts that Gunther von Hagens produces reflect the historical friction between scientific accuracy and artistic or aesthetic embellishment, which he does not perceive as conflicting. Each of his plastinates features a specific physiological feature (such as the muscular-skeletal, digestive, or cardiovascular-respiratory system) carved out with tantalizing precision; but what attracts the most

Fig. 8. Ganzkörperplastinate "The Runner," Bodyworlds. Gunther von Hagens, Institut für Plastination, D-69126 Heidelberg. www.bodyworlds.com.

attention are the artistic poses in which they are sculpted. As with the drawings in Renaissance anatomical atlases, we are diverted from the abhorrence of death and the cruelty of dissection by the lively appearance of each Ganzkörper. In line with the artistic tradition in anatomical drawings, von Hagens's plastinates are at least as determined by artistic conventions as by scientific insights. A plastinate called "The Chess-Player," showing the structure of the nervous system, stylistically resembles Auguste Rodin's bronze *The Thinker*. Another plastinated body, entitled "The Runner," demonstrates the workings of human kinetics. The fluttering bits of skin and tissue attached to its limbs suggest the dynamics of a running man,

Fig. 9. Umberto Boccioni, Unique Forms of Continuity in Space, *1913. Galleria d'Arte Moderna, Milan, Italy. Photo: Scala / Art Resource, New York.*

triggering associations with Umberto Boccioni's futurist art which represented movement and speed in new ways.

Despite the seemingly conflicting character of true-to-nature representation and artistic intervention, in Bodyworlds both are explicitly conflated in single bodies. Von Hagens's anatomical bodies, like those of his historical predecessors, incorporate contemporary artistic styles and convention—in his case, postmodernism. Although the full-body plastinates look like imitations of Renaissance anatomical art, they are modifications of earlier artistic styles. This is particularly illustrated by the "expanded bodies"—plastinates drawn out spatially to expose inner organs. One such expanded plastinate is stretched lengthwise, as a result of which the body looks like a sculpted totem pole or the inside of a Giacometti bronze. Another plastinate is expanded in four directions, the individual parts of the body suspended with invisible threads in the three-dimensional space thus created. A third is stretched horizontally from the diaphragm, revealing the intestines. Obviously, expanded bodies are not true-to-nature representations of the body; instead, they have been spatially reconfigured into giant three-dimensional models.

Fig. 11. Detail from Michelangelo's depiction of Saint Bartholomew, from the ceiling of the Sistine Chapel.

If this exhibition of plastinated bodies had simply carried on the age-old tradition in anatomical art of merging scientific soundness with artistic styles, there would have been no public outcry. But there is an important caveat to this alliance of traditional postmodern anatomical art. Von Hagens's sculptures are not *representations* of bodies, as are Vesalius's engravings in the *Fabrica,* or Govard Bidloo's copper engravings of the interior body in *Anatomia Humani Corporis* (1685).[29] The plastinates presented at Bodyworlds are *imitations of representations,* executed in modified organic material. The most explicit illustration of this is von Hagens's copy of Vesalius's muscle man, who poses carrying his own skin in his hand, as if he has just taken off his coat. A life-size reproduction of Vesalius's muscle man was depicted on the wall behind the "look-alike" plastinated model at the

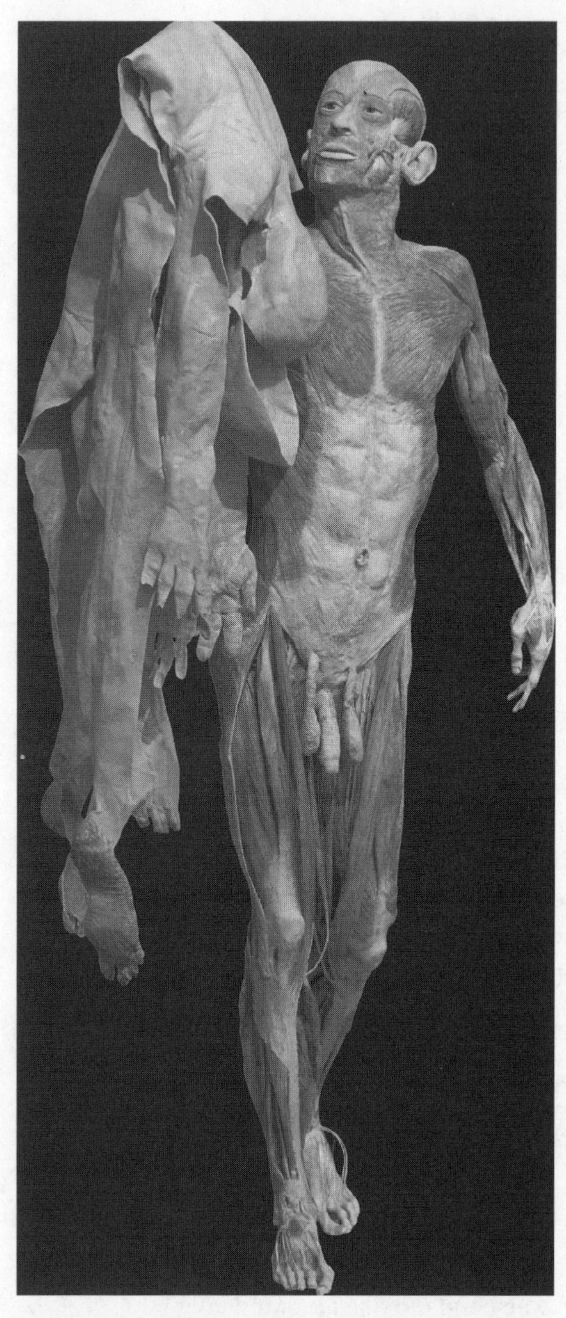

Fig. 10. Ganzkörperplastinate "Man with Skin on his Arm," Bodyworlds. Gunther von Hagens, Institut für Plastination, D-69126 Heidelberg. www.bodyworlds.com.

Mannheim exhibition. Because the "real" body plastinate imitates a piece of art—Vesalius's drawing—object and representation seem to fuse in the sculpted body. This plastinate, at first sight, looks like a "reanimation" of a representation, a wink to Renaissance artistic anatomical tradition.[30] Postmodern literature and art, of course, are full of such gestures—playful imitations of existing styles, known as pastiche.[31] But the "reanimations" smooth over an urgent question at stake here: can life—or, rather, death—imitate art? The plastinated copy of Vesalius's muscle man is created from an "authentic" body, which, because of its chemical modification, can no longer be called "authentic." In our "culture of the copy," observes cultural critic Hillel Schwarz, authentic and fake seem interchangeable, and their distinction is therefore obsolete.[32] Indeed, if the audience took offense at the fusion of artistic representation and organic materiality in the exhibited plastinates, von Hagens's actual reversal of art-representing-body into body-representing-art is, in fact, at least as disturbing.

Visitors to Bodyworlds were also treated to transparent cross sections of the body, in the form of the scheibeplastinate, or plastinated body slices. Body slices are not particularly new or shocking; crosscut bodies have

Fig. 12. Juan Valverde de Amusco, Historia de la composición del cuerpo humano, *1556.*

been showcased in anatomical museums for quite some time.[33] Yet, if we are to believe the German anatomist-artist, scheibeplastinates offer an unmediated look into the depths of the body, using the combined techniques of cryogenic cutting and plastination. A cryogenic saw cuts the deeply frozen body into thin slices, either horizontally (coronally), or lengthwise (sagittally). The slices look like two-dimensional representation, yet von Hagens fashions them into positions in which they ostensibly regain three-dimensionality. One of the sliced plastinates, titled "The Transparent Body," consists of eighty-three slices spaced some four inches apart, forming a reclining body almost three yards long. The plastinated slices seem to provide direct access to the smallest details of the inner body—complex structures that cannot commonly be perceived by the naked eye. According to von Hagens, the slices offer an entirely unmediated look inside the real body because "in today's media world in which people are increasingly informed indirectly, the need for unmediated, unadulterated originality is on the rise."[34] Just as the pioneers of photography hailed it as a mode of writing with light beams, so von Hagens hails plastination as a technique enabling direct inscription, eradicating all mediation between object and representation, rendering all subjective intervention—inherent to representation—obsolete.

Once again, von Hagens's techniques are not merely a perpetuation of age-old representational modes. Mechanical mediation—in this case the application of the cryogenic saw, various chemical solutions, color additives, and a plastic coating—inevitably transforms the appearance of the body. Like his nineteenth-century predecessors, von Hagens assumes the superiority of mechanical objectivity, thus perpetuating the myth of transparent scientific truth: a pure representation of the human body without the contamination of human intervention. But besides sustaining this myth, he implicitly reverses the roles of anatomical object and representation. As with his imitation of Vesalius's muscle man (a reanimation of a representation), his scheibeplastinate "copy" in actual flesh the most common medical-visual representation of the late-twentieth-century body: the MRI or CT scan.

The introduction and subsequent universal implementation of magnetic resonance imaging and computer tomography scans in medical practice has rendered the appearance of body slices fairly familiar. Images resulting from MRI or CT scans, sometimes cross sections less than a millimeter thick, actually look like slices of body parts, but of course they are "photographed" representations. These recent visual technologies have familiarized the public with the inferred relationships between such fragmentary, two-dimensional slices and the three-dimensional human body, even if the medical interpretation depends on complex insight and visual expertise on the part of the observer.[35] Von Hagens, in turn, restyles two-dimensional representations into three-dimensional organic sculptures. Since most viewers accept the implied relationship between the slices and real bodies, the unmediated naturalness attributed to plastinated body slices seems merely an extension of the MRI-induced gaze. In other words, it is precisely the familiarity of their appearance that distracts from the act of violence involved in cutting the body into slices, and makes us forget that these objects are "fleshed-out" scans.

It is difficult to properly evaluate contemporary anatomical art without the perspective of art history. Like Vesalius, whose anatomical illustrations were partially based on the conventions of ancient Greek sculpture, von Hagens looks to artistic conventions for inspiration. However, in order to comprehend their various signifying layers, the plastinated sculptures also call for some knowledge of representation theory. I argue that von Hagens does not deploy a "true-to-nature technique" but fashions his artifacts into a "true-to-technique" nature. What is remarkable about von Hagens's technique is not that he reverses the order between anatomical object and representation, but that he renders the very distinction between these categories questionable. We are urged to consider Vesalius's muscle man not as a representation but as the paper model for an animated representation, and because we are so used to seeing MRI scans as representations, we may not even flinch at seeing a plastinated body slice of a "real" cadaver. Bodies and body models, bodies and representations, seem to have become interchangeable in Bodyworlds. Plastinated organs, orthopedic cadavers, expanded corpses, and sliced body parts tell us that the anatomical body, which was already a mixed object of science and art, has also become a hybrid product of artistic models and modeled organisms. Just as the real tulip is now a tulip treated and perfected with chemicals to appeal to current taste, so the real body is now a cadaver that is surgically, chemically, and artistically modified in accordance with prevailing aesthetic standards. In order to understand the ethical implications of this technological imperative, it would be instructive to look at the settings of von Hagens's exhibitions—at the conjunction of the anatomist's technique and his preferred mode of display.

ANATOMICAL ART EXHIBITED

Anatomical objects—both bodies and models—have aroused the interest and curiosity of a large lay audience since the late fifteenth century. From the very beginning, anatomists have recognized the great potential for widespread professional publicity in exhibiting the process of dissection. In Vesalius's time, the anatomy lesson was a public spectacle; it was not until the late eighteenth century that cadaver dissection disappeared behind the closed doors of the hospital.[36] In some parts of Europe, the exhibition of body parts remained an attraction at traveling fairs as late as the early twentieth century. Between the anatomical theaters of sixteenth-century Europe and anatomical collections today, anatomical displays have gradually been assimilated into medical-scientific (museum) settings. Although the anatomical body never completely ceased to function as an entertainment and spectacle, it has slowly yielded to a more scientific-clinical gaze.[37] It is striking how von Hagens reinfuses the element of spectacle into the display—stimulating viewers' visceral identification with the dissected bodies. The physical setting of Bodyworlds, the self-created image of the anatomist-artist, and the provocation of the visitor all contribute to its popular status.

Von Hagens chooses unusual settings for his unsettling displays. In Germany, Bodyworlds was situated in neither an art museum nor a science museum but in the Mannheim Museum for Technology and Labor. The large industrial museum, with its rusty image, came to life when approximately ten thousand visitors per day swamped the corridors, lining up between steam engines and historical labor tools to see the plastinated cadavers. In Vienna, the anatomical bodies were not exhibited as elevated museum pieces. People stuck their noses into cadavers or chatted on cell phones while leaning on glass showcases containing body parts; medical students in scrubs circulated, providing explanations on demand, and the café was separated from the body sculptures by nothing more than a rope divider. At one point, two circus acrobats from the neighboring Cirque du Soleil gave a show amidst the plastinated cadavers, in order to "emphasize the remarkable similarities between the muscular structures of the corpses and the acrobats."[38] The exhibition effectively erased the distance between viewer and object that is common in traditional anatomical museums, and diminished the awe and fear usually felt vis-à-vis the dead.

Another teasing element integral to Bodyworlds is Von Hagens's self-presentation as artist-anatomist. Whereas in the Renaissance, anatomical art was commonly produced by a team of two professionals, each holding distinctive professional credentials, von Hagens stresses his dual scientific-artistic license. On the one hand, he presents himself as "professor doctor"; by constantly articulating his academic titles, he seems to want to establish himself as an anatomist. The Institute of Plastination, where he produces his plastinated cadavers, is supposedly affiliated with the University of Heidelberg (where von Hagens once worked), although it is privately funded. The academic status of von Hagens and his institute are firmly grounded in existing scientific networks in multiple ways. On the other hand, in his presentation and appearance von Hagens adopts the identity of an artist—of Joseph Beuys in particular. Friends and foes alike have compared Gunther von Hagens to the famous German artist and even called him *"die Leichen Beuys"* ("the cadaver Beuys"); this is not surprising considering the photos and videos in which the man in the white lab coat is wearing a Borsalino hat, one of Beuys's trademarks. Professor Doctor von Hagens fashions himself as the eccentric artist, whose stamp of artistic idiosyncrasy marks each of his plastinated objects. Perhaps von Hagens's mixed profile annoyed some visitors, but more than that, it may be the element of performance that made people uncomfortable. Not unlike Joseph Beuys, the artist-anatomist foregrounds the process of production (in his case plastination) rather than hiding it from the audience.[39] On a video that was projected on a large screen at the Mannheim exhibition, visitors could see von Hagens at work in his studio, submerging bodies in large basins of pink fluid.[40] With scalpels and knives, he cuts away body fat and chisels his object into shape, much like a sculptor, but also reminiscent of the violent dissections practiced in old anatomical theaters. At the London exhibition, held in a gallery, he even performed a public dissection, the first public anatomy lesson to take place in Great Britain since they were outlawed in 1832.[41] In front of 500 visitors, von Hagens, wearing blue overalls and his Beuys hat, transformed the body of a donated corpse into a pile of organs within three hours.

Renaissance anatomical theaters encouraged visitors to identify with the corpse on the dissection table; after all, the cadaver mirrored the observer's own living body and foreshadowed what lay ahead. The Bodyworlds exhibitions played on the audience's mortal anxiety, inviting them to engage in thanatoligical voyeurism and narcissistic identification. But Bodyworlds even took this process of identification one step further: Before leaving the exhibition, visitors could obtain forms to sign up as *Körperspender,* or future donors for plastination. To those wishing to donate their own body to science, plastination presents the opportunity to unite posthumous altruism and (more egocentrically) eternal "life." Since antiquity, people have sought to save their mortal bodies from total decay by having them mummified or embalmed after death. Yet, in contrast to embalmed cadavers, von Hagens's plastinates, since they are exhibited as anonymous figures, cannot be traced back to a specific person.[42] In the guise of helping the progress of science, plastination offers the opportunity to preserve one's mortal remains for centuries, if not longer, and to be displayed in all one's naked grandeur in front of millions of curious, astonished, and fascinated observers. It is perhaps not surprising that many of the prospective donors, when asked on the application form about their motivation, claimed to be elated at the prospect of having their body transformed into a statue or work of art.

Bodyworlds systematically fed on two ways of looking—scientific and artistic—that are traditionally bound up in anatomical modes of display, and also added an element of popular engagement and corporeal identification, reviving its roots in the tradition of public anatomy lessons. Setting, artist's image, and audience appeal all gave a sensational edge to the ostensibly sanctified artifacts commonly treated with awe and respect. Von Hagens's transgressions of boundaries severely tested the visitor's ability to switch contextual frames, and the anatomist-artist often invoked historical traditions of public dissections and anatomical theaters to justify his practices. Indeed, it was exactly the artistic and entertaining character of the exhibition that became the center of criticism.

THE PLASTINATION CONTROVERSY

The media debate triggered by Bodyworlds revolved around the ethicality of the exhibition and was discussed almost exclusively in binary terms: art vis-à-vis science. The most outspoken objections came from moral theologians who were offended by von Hagens's desecration of human cadavers; they respected the scientific use of anatomical bodies, but loathed the artistic motives of the "anatomist." A second vocal groups of opponents were medical scientists. Anatomists and other medical specialists, who usually subject themselves to rigid protocols regulating the donation and treatment of corpses, strongly objected to von Hagens's violation of these ethical norms in using dead bodies for frivolous purposes; in this case, for art. Directors of Europe's leading anatomical museums responded negatively to the Mannheim exhibition.[43] They declared that plastination added no scientific or educational component to body models. Moreover, they resented von Hagens's sensationalism and his catering to the art world; anatomical exhibitions should be in the service of science, they argued, not art. In interviews with visitors, reactions ranged from enthusiasm and fascination to indifference and strong ethical objections to the use of human cadavers for other than scientific purposes. Such dubious practices, some contended, would raise eyebrows anywhere in the world, but particularly in Germany, where historical anatomical tradition is imbued with Germany's tainted history involving scientific experiments on living and dead human bodies and the Nazi ideology of eugenics.

In response to his many and powerful foes, von Hagens took a surprising defensive position: understating its artistic value, he emphasized the strictly scientific nature of his work. For instance, in Mannheim he hired medical students to provide information to visitors and to answer their questions. The donor forms distributed at the end of the tour resembled official organ donation forms. Countering criticism from the medical profession, von Hagens underscored the academic status of the Heidelberg Institute for Plastination, as well as his own scientific qualifications. The anatomist-artist claimed that he was trying to liberate science from its ivory tower; unlike his peers, he considered the education of a broad audience an important asset to the discipline. According to von Hagens, the success of Bodyworlds proved that ethical norms are no longer imposed from the top down (for instance, by church authorities) but that individuals themselves decide what they consider ethical or not. While it is understandable that von Hagens, for pragmatic reasons, would capitalize on his credentials as a scientist to defy his critics, it seems nonetheless peculiar that he would fall back on a professional hierarchy that privileges the scientist at the expense of the artist, while, historically, these categories have always been blurred.

Both the opposition and von Hagens's defense reinforced a false dichotomy between art and science, implying that a different set of ethical norms or standards applies to each. I don't think that is true, however. Although the average contemporary-art-exhibition visitor is expected to be more "shock-proof" than the average anatomical-science-museum visitor, the use of severed limbs or body parts in a work of art would definitely cause unease and thus probe ethical norms. In recent years, we have witnessed various modes of artistic expression that, deliberately or unintentionally, have questioned the sanctity or integrity of human flesh. Some sensational exhibits, featuring organic tissue, have pushed the envelope of ethical permissibility in the art world.[44] It is clear that von Hagens walks a tightrope between artistic articulation and ethical judgment. When asked about his aesthetic-ethical sensibility, he invariably invokes historical tradition to justify his blending of scientific tools and artistic styling. There is no denying the aesthetic refinement of his sculptures; however, their naked sensationalism is every bit as attention-grabbing as the work of some contemporary artists or, for that matter, scientists. The sensationalism, exploitation, and blatant commercialism characterizing Bodyworlds unequivocally exposed von Hagens as a savvy businessman.

The most disturbing aspect of von Hagens's plastinates, in my view, is neither the transgression of artistic-scientific boundaries that purportedly fuelled the controversy nor the resuscitation of public spectacles. An idea that remained virtually untouched in the public debate concerning the exhibition was that plastinated cadavers prompted visitors to reconsider the status and nature of the contemporary body, both dead and alive. The contemporary body is neither natural nor artificial but the result of biochemical and mechanical engineering; prosthetics, genetics, tissue engineering, and the like have given scientists the ability to modify life and sculpt bodies into organic forms that were once thought of as artistic ideals: models and representations. What von Hagens does with dead bodies is very similar to what scientists and doctors do with living bodies. The reversal of body and representation also underlies the principles of cosmetic surgery; and the problem of authenticity and copy seems more urgent in genetic-engineering experiments than in plastination. Perhaps most unsettling about von Hagens's plastinated cadavers is their implicit statement that the very epistemological categories that guide us in making all kinds of ethical distinctions simply do not apply here. Categories such as body versus model, organic versus synthetic/prosthetic, object versus representation, fake versus real, and authentic versus copied have become arbitrary or obsolete. Since we commonly ground our social norms and values in such categories, von Hagens's anatomical art seems to elude ethical judgments.

Plastination is an illustrative symptom of postmodern culture, just as Frederick Ruysch's anatomical objects were a symptom of Vanitas art and Renaissance culture. Cadavers have become amalgams of flesh and technology, bodies that are endlessly pliable, even after death.[45] Bodies, like tulips, are no longer either real or fake, because such categories have ceased to be distinctive. Modified tulips that last longer and look absolutely perfect raise the same ethical and philosophical concerns as genetically engineered sheep and manipulated corn. By the same token, our gaze and our view of the body are increasingly challenged and cultivated by the plasticity of technology. In principle, von Hagens's sculptures might provide an interesting critical perspective on the all-pervasive influence of technology and the waning integrity of the flesh, yet his defense of plastinated cadavers divulges no such intentions. Rather than commenting on the confusion of boundaries, he reconfirms binary labels by promoting his cadavers as real, intact, and authentic. While obviously mixing scientific insights with artistic styles, he quickly reverts to the solid bastion of science when defending himself against criticism.

Paradoxically, von Hagens's denial of distinct categories of embodiment, and thus the ethical norms in which they are cemented, does not prevent him from invoking those same norms to claim the legitimacy of his practices. The artist-anatomist's plastinated cadavers seem exemplary of a culture that, according to

N. Katherine Hayles, is "inhabited by posthumans who regard their bodies as fashion accessories rather than the ground of being."[46] The culture of the posthuman is a continuation of the liberal/humanist tradition in which the body is regarded as a mere container for cognition, and the religious tradition that sees the body as a temporary vessel for the soul. Von Hagens attempts to detach his plastinated bodies from their living signifieds, yet they are inescapably infused with historical, local, and cultural meanings. Dr. von Hagens is perhaps best seen as a postmodern Dr. Tulp—who deploys medical technology to express a potentially provocative, but in actuality disturbing, commentary on our technological culture.

CHAPTER 4

FANTASTIC VOYAGES IN THE AGE OF ENDOSCOPY

IN THE 1966 SCIENCE-FICTION MOVIE *FANTASTIC VOYAGE*, A CREW OF THREE MEN AND A WOMAN (RAQUEL WELCH) EMBARK ON A SPECIAL MISSION.[1] THEY ENTER A SPACE CAPSULE THAT IS THEN SHRUNK TO MINUSCULE SIZE BEFORE BEING INJECTED INTO THE VEIN OF AN ANESTHETIZED PATIENT—A FAMOUS SCIENTIST WHOSE LIFE IS THREATENED BY A BLOOD CLOT. THE BODY VOYAGE LEADS THE CREW PAST AND THROUGH VITAL ORGANS, SUCH AS THE LUNGS AND THE HEART, CAUSING THEM TO MARVEL AT THE WONDERS OF corporeal space. "Man is the center of the universe. We stand in the middle of infinity to outer and inner space and there is no limit to either," whispers Mr. Grant, one of the bionauts, in awe of the cosmic sights unfolding around the spaceship. After various suspenseful run-ins with organic obstacles, the miniaturized team finally reaches the brain, where they successfully destroy the blood clot with the help of laser beams. They manage to escape from the body through his eye, washed out in a tear, just before growing back to ordinary size. *Fantastic Voyage* features a combined scientific and surgical expedition: the crew wants to see the body from within, convey images of its interior to the outside world, and repair mistakes in its organic fabric.[2] Richard Fleischer's movie is certainly emblematic of doctors' age-old desire to discover the body's interior secrets, chart its unknown territory, and correct its flaws.[3]

What, exactly, is the broader significance of this mythical journey aimed at exploring and fixing the interior body without leaving a trace on its outer coating? Is it an innocent fantasy? A utopian desire? A scientific projection of the future? I argue that the "fantastic voyage," including its underlying medical logic, actually functions as a trope that underpins the development, implementation, and dissemination of a major medical imaging technology: the endoscope. Derived from the Greek *endo* ("within"), and *skopein* ("to view"), the term signifies the power to extend the human eye to the body's interior. In its earliest forms, the endoscope did just that: it allowed doctors to peer inside the human body. Gradually, during the course of the twentieth century, endoscopy was no longer restricted to amplifying a doctor's ocular capacities but also aimed at extending a surgeon's manual skills under the skin, so that examination and operation could take place during one and the same surgical expedition. With today's video endoscopic equipment, doctors are able to visualize, diagnose, and remove unwanted growths inside the body, leaving only a small scar on the patient's skin. Technological progress has reached a provisional apogee with the latest development of "virtual endoscopy," a technique that will purportedly turn science-fiction fantasy into surgical reality: surgeons predict they will soon be able to move around and operate inside a body without penetrating the skin.

The development of surgical and diagnostic instruments never occurs in isolation, but is part of an intricate web in which technology, medical practice, and cultural representation are mutually constitutive. Regarding endoscopy, we see how its development is inextricably bound up with innovations in media technologies, such as photography, video, and computers. Endoscopic techniques have changed medical practices dramatically—particularly the skills of the surgeon, the setting of the operating theater, and the involvement of the patient. Double-serving as a media technology, endoscopy has also profoundly affected the ways in which the interior body is conceptualized and represented in popular media such as television. The interplay of technology, medical practice, and cultural appropriation becomes manifest in what I will refer to as "the endoscopic gaze." In contrast to Laura Mulvey's "cinematic gaze," the endoscopic gaze signifies the surgeon's view *from within* the body, enabled by medical technology.[4] In the case of the cinematic gaze, the spectator is forced to adopt the perspective of the camera, which often produces a depersonalized, detached view of the object; the endoscopic camera similarly creates a seemingly "objective," neutral view of the object, in this case, the body's interior. In the past fifty years, our perspective on the interior body—mediated by the surgeon's eye—has shifted from outside to inside. We no longer peer from the outside in, through an incision in the skin; instruments now allow us "immediate" access to the body's tiniest details. This insider's point of view vis-à-vis the body has, meanwhile, pervaded our culture.

But what are the implications of this ubiquitous endoscopic gaze? How do innovations in endoscopic techniques affect doctors' and patients' concepts of corporeality? And do such innovations, in turn, influence our collective appreciation of surgical intervention and its consequences? In other words, do they redefine our standards and expectations concerning the perfectibility of the body as a physical container? Between the earliest trials with endoscopes and the latest experiments with computer-based visualizing technologies, the myths of transparency and nonintervention have transpired equally in medico-technological developments and in their clinical and cultural adaptations. Sketching the past, present, and future of one particular imaging instrument, I argue that, in the conceptualization of the permeable body, visualizing technologies are inevitably bound up with our visions of technology.

LOOKING FROM THE OUTSIDE IN

A primitive form of endoscopy emerged in 1800. The earliest visualizing aids used by physicians to look inside the body resembled telescopes or binoculars. The gynecological speculum and the rectoscope, based on the reflection of light from the outside, allowed physicians to view internal cavities and interstices between organs through natural openings. In the 1850s the first "gastroscopes" were introduced, allowing more internal areas, like the stomach, to be reached by passing a straight, static tube through the esophagus.[5] In 1877, the German physician Max Nitze expanded the field of vision by bringing artificial light sources inside the body; his "cystoscope" carried a platinum wire heated to whiteness into the bladder via a tube, followed by a sort of telescope. He later replaced the platinum wire with electrical wire of the Edison lamp. In the early twentieth century, surgeons began to use endoscopy not just to enter the body via natural openings but also to enter it through purposely made incisions in the skin. Percutaneous techniques, as these procedures are called, are inherently intrusive, as they require surgery and anesthesia.

Until the mid-twentieth century, endoscopies remained highly invasive procedures; quite often they were surgical interventions, whose potential benefits hardly outweighed the dangers. If a surgeon decided, either during or after endoscopic diagnosis, to perform additional surgery, the entire setting of the operating theater was fashioned to suit the surgeon's view, looking from the outside in. Magnifying glasses, forehead mirrors, and lamps aided the surgical gaze. During such open surgical procedures—still the most common form of surgery—the hands of the surgeon are in direct contact with the patient, and the incision in the skin must be wide enough to allow the surgeon's fingers or instruments to touch the diseased body parts. As German sociologist of medicine Stefan Hirschauer eloquently puts it in his "thick description" of the surgical act: "One must see to cut more and cut to see more."[6] The patient's body is reduced to an operating area, a silenced object, veiled except for the sterile opening in the skin. Anesthetized and unconscious, the patient becomes a virtual participant, talked about but not talked to; anesthesia and/or cloth screens protect the patient from pain, but also from unpleasant views, fear, and shame of his own naked interior.[7] The operating theater itself is a delimited area. In the sterility zone, the body is shielded from bacteria and germs, the walls mark off the domain of privacy, and entering the operating theater is tantamount to entering the patient.

To render endoscopy more useful and less invasive as a diagnostic device, three main requirements had to be met: sophisticated optical equipment, a safe and reliable internal light source, and, preferably, a flexible cable which could be easily passed through spaces inside or between organs. In 1957, the fiber-optic cable, invented by the Danish engineer Holger Møller-Hansen and later perfected by the English American surgeon Basil Hirschowitz, offered a solution to the basic problems of optics, light, and tube flexibility.[8] The new cable consisted of thin Perspex threads that transmitted light from fiber to fiber; it was fully flexible and could be inserted through most natural and artificial openings without losing optical quality when bending the tube. A magnifying lens was placed at one end of the cable and an eyepiece at the other. The new cable considerably improved the surgeon's visual scope within the body; more importantly, it increased the patient's comfort and safety.

It was not until the 1960s that technicians managed to combine photography with endoscopy, enabling doctors to bring back souvenirs of their exploratory missions. Almost concurrently with the advent of the fiber-optic cable, in 1963 the Japanese company Olympus developed a mini-camera that could be attached to the end of the cable. Equipped with a wide-angle lens, flashlight, and film cartridge, doctors could now take usable snapshots of interior landmarks and landscapes from *within* the body's interior; the skin had become permeable to the camera's eye. A major technical problem at the time was that the still pictures had to be taken "blindly" because the lens could not be steered from the outside. Moreover, the wide-angle lens inherently distorted the picture, and focus was often disturbed by the presence of bodily fluids such as bile, blood, or mucus in front of the lens.

Although endoscopic pictures played a minor, supportive role in surgery up until the 1970s, they proved very useful in spreading the endoscopic gaze outside the medical domain. In the 1960s and 1970s, medical journals began to reproduce endoscopic pictures—particularly as evidence of a successful intervention. "Before and after" images showing the removal of pathological growths or foreign objects were silent witnesses of surgeons' pioneering professionalism. Deploying cameras, doctors could produce impressive evidence not only of their ocular presence inside but of their missions accomplished. Early endoscopic pictures also served as teaching aids, familiarizing future surgeons with the endoscopic gaze. As photographic quality improved and color photography became commonplace, the intriguing snapshots continued to disseminate this gaze, exposing the world beneath the skin to the public eye. First medical journals and then popular magazines were eager to publish surgeons' awe-inspiring bodyscapes. *Time, Ebony,* and other magazines began featuring vistas the general public had never seen before: duodenal ulcers, infected vocal membranes, or a ruptured meniscus.[9] The organs in photographs were rarely identified as belonging to individual persons; rather, they signified generic healthy or diseased body parts, protecting the privacy of the person portrayed. These pictures of the living body's interior helped the general public imagine what it was like for a surgeon to look from the outside in. Yet these pictures were as exotic as the first pictures from the moon.

TURNING THE INSIDE OUT

With the mass distribution of endoscopic images, made possible by the introduction of video techniques in the early 1980s, the inner body became less exotic and more familiar. A small video camera attached to a television cable replaced the fiber optic cable; inserted into the body and steered from the outside by the surgeon, the camera sends digital images to a video screen in the operating room.[10] Unlike the photo camera, the video camera sends live moving images of the body's interior—pictures that can be magnified and manipulated outside the patient's body. Most importantly, video endoscopy extends not only the eye but also the hand of the surgeon: through the same tube that allows the video camera into the body, the surgeon can insert operating instruments. Pure diagnostic use of endoscopy gradually gave way to combined diagnostic-therapeutic surgery, just as featured in *Fantastic Voyage.*

Almost twenty years after its implementation, video endoscopy is performed in many medical specialties, ranging from gastoenterology to gynecology, from urology to orthopedic surgery. Therapeutic uses include the removal of foreign objects, tumors, and kidney stones, and the insertion of drains in obstructed bile ducts. Kidney and bladder stones may be crushed endoscopically by means of an ultrasound laser, or entire stones can be extracted through the tube. As many as 60 to 80 percent of all gallstone removals are now performed with the help of endoscopy.[11] The advantages of video endoscopy are widely acclaimed. The procedure is less invasive and thus less traumatic to patients than open surgery. It can sometimes be performed under local anesthetic, resulting in quicker recovery times and economic savings. And there are the cosmetic advantages. Whereas, in the case of open surgery, the removal of a gall bladder requires a six-inch incision in the abdomen, a video-endoscopic procedure leaves only the barely visible marks of a few stitches at the edge of the navel.

To a layperson, video endoscopy creates a strong sense of the body's transparency, yet for surgeons this technology means less, rather than more, access to the patient's inner layers. The surgeon is no longer looking directly at the insides of a real body, its organs and intestines laid open through an incision in the skin, but at a *mediated* body—mediated by the camera and video display hanging over the operating site. Endoscopic surgery requires radically different skills from open surgery: hand-eye coordination occurs through the video viewer and its electronic display. Because the patient's skin is only minimally pierced, the surgeon has to navigate her instruments through minuscule openings. Whereas open surgery aims at

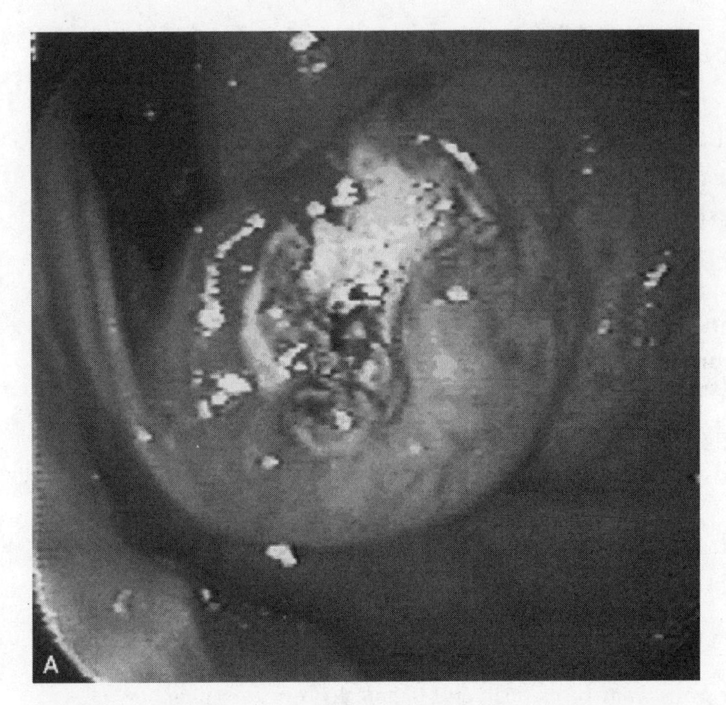

Fig. 13. Endoscopic image made during operation. Courtesy of Academic Hospital Maastricht.

enhanced visual acuity in the "original material" itself, the endoscopic gaze tries its utmost to overcome the local circumscription of the eye. Mediation through the video endoscope may limit the surgeon's view to such an extent that she misses vital information. For instance, after the endoscopic removal of stones in the gall bladder, chips that are invisible due to the instrument's limited scope may obstruct the entrance to the liver or the pancreas, sometimes leading to additional open surgery to rectify complications arising from the instrument's limitations.[12] Every instrument that opens up new vistas also sets new restrictions. Despite its claim of visual transparency, the endoscopic gaze provides an inherently constrained perspective on the interior body.

From the patient's point of view, by contrast, video endoscopy seems to increase the transparency of the body. In the case of open surgery, the patient is usually unconscious and veiled by a cloth screen, preventing him from witnessing the procedure. While most types of video endoscopy are still performed under complete anesthesia, an increasing number of procedures require only local anesthetic, allowing patients to watch their own operations. Lying on the operating table, a patient can keep track of the surgeon's actions by watching the television monitor. It seems much easier to watch such an operation via the screen, as it somehow distances the patient from his own body and protects him from the direct view that may trigger feelings of fear or shame. Because there is no pain, there is no visceral connection between what the patient sees on the screen and what is happening to his body. Because of the procedure's exotic visuals and low level of messiness, the patient may even become intrigued by his own operation. In cases where endoscopic surgery is performed under full anesthesia, the patient may take home a videotape; some scholars claim that retrospectively watching one's own operation may even have therapeutic effects.[13]

Video endoscopy not only renders the interior body visible to the patient but also invites the gaze of outsiders. Video recordings may be used to train medical students or interns. Patients may wish to take the tape to another specialist for a second opinion.[14] And, needless to say, from the very beginning, recordings of endoscopic procedures have been used as evidence in medical malpractice suits.[15] The same instrument that allowed the transposition of the endoscopic gaze from the operating theater into the courtroom also brought it into the living room. Before the 1980s, medical professionals considered the presence of television crews in the sterile operating environment intrusive. Surgical acts were occasionally filmed for a general public—open wounds, bloody intestines, and the gloved hands of the surgeon offering a less-than-pleasant view. Yet the advent of video endoscopy radically changed the relative intimacy of the patient's inner body and the operating theater. Since endoscopic pictures look less bloody and messy than images of open wounds, they are more enticing to a general audience.

In recent years, endoscopies have been regularly featured on European public television stations, either in science documentaries or in health programs, familiarizing the public with the fascinating view from within. *Surgeon's Work* is a typical example of a popular medical talk show in the Netherlands.[16] Specialists and patients sit in a studio, where a moderator introduces patients whose diseases are used to demonstrate the latest and most impressive surgical techniques and instruments. One program in the series, titled "Body Voyage: Endoscopic and Percutaneous Techniques," showcases a variety of endoscopic procedures, including a bronchoscopy to clear the sinuses, a laparoscopic inspection of the ovaries and Fallopian tubes, and an endoscopic kidney stone removal. At the beginning of the program, the moderator invites his viewers on "an exciting voyage through the human body." The patient is introduced, a man who had gone to see the urologist (sitting next to him) for what the moderator calls a "typical old man's disease": prostate enlargement. The "old man," visibly shy and occasionally giggling, lists his symptoms—pain in the urinary tract, trickling urine—after which the surgeon translates this common problem into medical jargon. Next, the camera takes us to the operation, filmed about a week earlier. Without any transitional shot, we are taken straight into the man's body via the camera in the uroscope; the surgeon's mechanized eye guides us through the urinary tract. Meanwhile, the

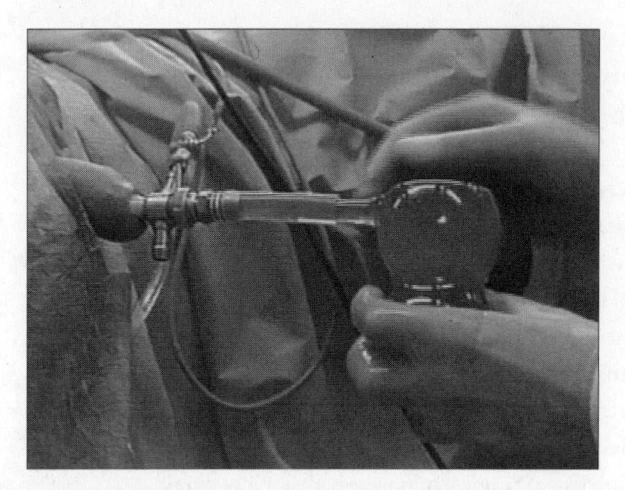

Fig. 14. Still from Surgeon's Work. (Chirurgenwerk). *Evangelische omroep, Netherlands.*

surgeon comments on the safety of this highly advanced technique. With the help of a pulse-regulated electric wire, the surgeon removes the surplus cell tissue that hinders the passage of fluids through the tract, and cauterizes the open blood vessels. Every few minutes we get a wide-angle shot of the entire setting: the patient's anesthetized body is completely covered by blue sheets, except for the top of his penis, while the urologist sits on a stool between his legs. Only at the very end of this segment, when the doctor demonstrates how to flush and clean the urinary tract, do we get a glimpse of the instrument penetrating the body through the natural opening. Just as an incision in the skin repels ordinary viewers, so would the actual penetration of the penis by the uroscope. Back in the studio, a buoyant, smiling patient assures viewers that he has suffered no pain or adverse consequences from the operation; shyness is replaced by admiration for the surgeon.

The scene is not unlike a guided tour through a coal mine, where one can watch as sophisticated technology is deployed to open up a caved-in tunnel. The round edge of the tunnel limits the viewer's gaze to a circumscribed focus; dim light adds a touch of suspense to the locus of action. Through the video camera, which registers the surgeon's gaze as well as his acts, the audience only gets to see what the tour leader wants it to see: advanced technology, a skillful surgeon, and a satisfied patient. Mediated by the endoscopic camera, the skin that "personalizes" the man, while remaining literally intact, is stripped away symbolically. The endoscopic camera coincides with the television camera, enabling viewers to visually travel through the man's penis and urinary tract. Although we do not see any part of the man's skin during the surgical procedure, we are able to connect the moving images of the interior body to the patient in the studio, identifying the owner of this male reproductive organ. Peculiarly, exposure of the man's insides seems less reason for public shame or embarrassment than the filming of his penis from the outside. The potentially voyeuristic, sexually connoted gaze, directing the eye from the outside to the inside of the body, is completely taken over by the clinical-endoscopic gaze, the view from within that evidently effaces all nonmedical signifieds.

Almost imperceptibly, various boundaries are transgressed in this program. The television talk show is formatted as a *mise-en-abîme.* While sitting in our living rooms, we watch the people in the studio watching themselves on video in the operating room, where the operating surgeon takes us on a journey through the keyhole of the endoscope into the belly of the patient. Despite switching between levels of representation, we tend to forget about the boundaries that we (tres)pass: the television screen, the walls of the studio, the operating theater, and the patient's skin. Operating room and television studio are no longer separate spheres, but merge on the screen in our living room. In the cultural deployment of clinical endoscopy, the surgeon becomes a crew leader, the television camera a vehicle for inspection, and the patient a corporeal universe. The once-private inner body has been transformed into a public sightseeing space, and, as "fantastic voyage" logic dictates, is innocent, informative, and even fun to watch. But just how innocent or informative is the transgression of boundaries in programs like this one?

Minimally invasive surgery undeniably has great advantages for the patient; yet moving inside and outside the body's interior now seems so easy, effortless, and painless that there is a serious risk that surgery will become more common.[17] Whereas once the diagnostic was more threatening than both the disease and the remedy, now the relatively low impact of the combined diagnostic-operative tool may mean that mild symptoms of stomach upset or pain in the abdominal area may lead to all-too-easily-performed medical imaging procedures.[18] In the public mind, an endoscopy may be equated to an X ray or ultrasound, disguising the inherent invasiveness of each of these procedures. But endoscopies are still surgical interventions, and even if they leave only minor scars on the skin, every penetration of the surface is still an invasion of its interior.

Televised body voyages foster false ideas about transparency and nonintervention. Although they claim to give viewers the experience of being "live" witnesses of actual surgical expeditions—uncut, straight from the locus of action—these journeys are highly mediated by dramatic conventions and broadcast protocols. The sense of immediacy is largely the result of optical effects, rather than of technology per se. Editing techniques are structured by narrative conventions, and the sequence is determined by dominant aesthetic and ethical standards. Obviously, corporeal space is carefully retouched to show only appealing views, omitting any scenes that would repel the viewer. Televised body voyages suggest that the surgical expedition is a harmless operation, as evidenced by a virtually untouched skin; by the same token, televised scenes protect the viewer from feeling too much empathy, discomfort, or unpleasantness. The dominating clinical gaze in these programs is the result of a double mediation: the body is filtered and cleaned up by both the endoscopic camera and the television camera. Surgery becomes a noninvasive, almost aesthetic experience, unadulterated by pain, scarring instruments, or potential complications. The new video-endoscopic gaze profoundly affects the configuration of inner space, leaving the impression that imperfections or undesirable growths can easily be removed by noninvasive technology, leaving the body untainted.

LOOKING FROM THE INSIDE IN

In 1966, *Fantastic Voyage* gave its viewers an imagined glimpse of a future in which operations without knives and scars would be common surgical practice. Thirty-five years later, it is not just filmmakers who entertain the fantasy of experts traveling in the body, removing life-threatening obstructions, and escaping through natural openings. Medicine has generated its own futurologists, and it should come as no surprise that all-too-frequently the medical imagination is fed by the same mixture of projection and extrapolation that has inspired Hollywood directors. In the past decade, surgeons and medical experts have publicly prophesied how more advanced computer techniques will increase our current potential to visualize, travel inside, and fix a diseased body. The published projections of a few prominent medical scientists, whose extrapolations of current technological developments betray a distinct belief in the ultimate fantasy of the perfect, permeable, and untainted body (and whose fantasies may well steer the future of medical technology and practice), deserve closer scrutiny.

At the onset of the twenty-first century, digital techniques are rapidly replacing analogue instruments. The growth of computer-assisted imaging technology began in the early 1990s, when digital pulse modulation, or digitization, made telecommunication systems (television and video) fully compatible with medical visualization instruments, such as computer tomography, magnetic resonance imaging, and confocal microscopy. These instruments allow three-dimensional viewing into the body, which should help overcome serious constraints imposed by analogue video endoscopy. Video endoscopy's mediating camera and screen render a flat representation of three-dimensional organs and interstices. In addition, the camera, wandering inside the body, allows only a single perspective, restricting the focalizing capacities of the surgeon. Although the video-endoscope gives access to the body under the skin, this access is limited: the camera cannot actually go inside the heart, the brain, or the kidney. And even though there are now mini-cameras as small as half a millimeter in diameter that can be inserted into arteries to detect blood clots, the video cannot register the inside of a cell. Finally, the discomfort caused by (local) anesthesia and the incision—even though this is minimal—still imposes undesirable stress on the patient.

Digital or virtual endoscopy, according to medical futurologists Richard Satava and Richard Robb, will emulate video endoscopy in many respects and, ultimately, render the body transparent.[19] Unlike "real" endoscopy, its virtual equivalent offers three-dimensional representations, displayable from every possible angle, allowing full optical access even into the most hidden and minute parts of our body. Computers can reconstruct three-dimensional images from digital data obtained by various body scanners. Coupled with sophisticated computer algorithms, these data render high-resolution images that can imitate, and factually emulate, the moving films of an endoscope. This enables the surgeon to visualize the interior body from virtually every plane, to penetrate the walls of organs, and to fly through ducts and tracts where it is physically impossible for a camera to travel. The trick of virtual endoscopy is that it resembles a video film taken with an actual camera inside the body, yet, in fact, it is a projection of that interior, extrapolated from digital data. The real body is represented as spatial information, resulting in a high-resolution visualization that is neither a photo nor a model, but an animated reconstruction of computed data.

Future digitization of the operating theater, including the use of virtual endoscopy, will substantially affect surgical practice. Satava and Robb predict that in the not-too-distant future surgeons will be able to walk around in a virtual body. Rather than the flat, two-dimensional video images displayed on a bulky screen in the operating theater, the images resulting from three-dimensional virtual endoscopy can be spatially displayed in three ways. The surgeon could wear a head-mounted display or stereo glasses that strategically position her inside the obtained data; moving around in a virtual body places the user inside the visualization domain and allows the surgeon to look inside from within.[20] Satava also envisions surgeons walking around in an eight-foot-square room where three-dimensional images are projected.[21] More suitable for instruction and publicity, yet completely useless for diagnostic purposes, would be the display of digital-physiological data as an animated video sequence on a screen, following a predetermined flight-path—for instance, through the esophagus or the bronchi. Technicians can modify algorithms to such an extent that animated image sequences or fly-throughs appear as realistic spatial movements on the screen. In other words, virtual endoscopy offers viewing capabilities that exceed the potential of real endoscopy, so the surgeon's eye and the public eye can travel to places that the material camera cannot. Virtualization of the patient's body is, in fact, the reversal of the idea of miniaturization as pictured in *Fantastic Voyage:* rather than shrinking the surgical crew, we inflate the body to disproportionate size.

In the future, this technology should help the surgeon navigate and operate. Satava predicts the development of operation robots who will perform surgery much more accurately and smoothly than real surgeons, and he foresees an increased use of laser beams, which may remove unwanted growths without penetrating the skin. Naturally, these new technologies will profoundly alter surgical skills. Perception, navigation, and cutting will all become computer-mediated activities, surgical acts performed via joystick or computer mouse. A new generation of surgeons, Satava posits, needs to be trained in "decoupling their oculo-vestibular axis for visual orientation from their haptic-proprioceptive axis for manipulating."[22] Children who have played computer games all their lives have a natural ability to uncouple visual and manual skills, as they regard the mouse as an actual instrument of navigation. Today's Nintendo kids, Satava assures us, will be tomorrow's digital physicians—"those for whom it is natural to look at a video image as a real object and who can use a joy stick as a pair of scissors."[23]

In this vision of future cybersurgery, a patient's body is more a collection of data than a physical object. Moving around in corporeal space, the surgeon can merge scanned data from the body with anatomical or standardized images to accurately identify and localize aberrations. The actual proximity of surgeon and patient is no longer relevant. Virtual endoscopy and computer-assisted visualizing technologies open up the way for "telesurgery," whereby the computer obliterates the physical distance between surgeon and patient.[24] Due to high-speed data transmission, the patient can lie in Tahiti while the surgeon operates in Siberia. These developments will quite literally turn the operating theater into a computer room or control center—an image reminiscent of *Fantastic Voyage.*

The utopian concept of surgical intervention that has no impact whatsoever on the actual body figures prominently in medical futurologies. The painless and traceless surgical-diagnostic expedition, as heralded in Fleischer's movie, seems to be the Holy Grail of cybersurgery. Richard Satava envisions computed visualization to be the ultimate diagnostic-operative tool that renders a medical checkup as easy as a routine walk through an airport scanner:

> A patient enters a physician's office, and passes a doorway, the frame of which contains many scanning devices (like airports today) from CT to MRI to ultrasound to near infrared and others. These scanners not only acquire anatomic data but also physiological and biochemical information. When the patient sits down next to the physician, a full 3-D holographic image of the patient appears suspended above the desktop, a visual integration of information acquired just a minute before by the scanners.[25]

In this utopian projection, the body reconstructed from digital data will hold no more optical secrets for the physician, who can diagnose disease and plan intervention. Cybersurgeons may try out some virtual bodily mutes in order to assess various potential effects of different interventions and, on the basis of that information, make informed decisions. The body arrested in digital space, made amenable to the virtual time of computed data, is pictured almost like a "tentative body"—a "fantastic body" that can be manipulated at will by the physician. The scientist can roam freely, discovering the body's ultimate secrets. Actual bodies are remarkably absent in these futurologists' visions; they seem merely containers or receptacles for the virtual examination's outcome.

Intervention through nonintervention is hardly a new promise, having surfaced as a literary trope in the annals of Western civilization for many centuries; rather, it is an old promise inscribed in new technology. By no means do I want to minimize the virtues of virtual endoscopy for surgical education and training; however, these medical futurologies challenge and complicate our ability to distinguish between reality and fantasy, projection and extrapolation. Currently, virtual endoscopies are only used experimentally as diagnostic tools, or as instructional models. Long-term clinical trials, recently started by the Food and Drug Administration (FDA), must prove the actual value of virtual over video endoscopy and the reliability of computed images.[26] The technical validity of digital endoscopic pictures is far from self-evident: How accurate are computer algorithms that translate data into images? How do they distort them? And what is the precise relation between video-endoscopic images and their virtual equivalent? The question is no longer whether the pictures are sharp or bright enough to see, but whether they actually show what is there and do not show what is not there.[27] The accuracy of virtual endoscopy is hard to establish, since its reference point—video endoscopy—is itself mediated by an instrument, and is thus a representation. A digital picture is already an intervention or, as cybertheorist William J. Mitchell calls it, a *mise-en-image:* a combination of computation, projection, and representation.[28]

Because of the highly realistic quality of digital images, virtual endoscopy has been quickly incorporated into popular culture. Scanned data are much easier to manipulate and integrate into other visual environments than are video images. Virtual body traveling has become the newest attraction in visual entertainment. *Body Story,* a British educational docudrama, features a different journey through the body in each part of the series.[29] Alternating enacted scenes and "reanimated" sequences of virtual endoscopies, the program creates the illusion that viewers are witnessing the onset and progressive development of a heart attack. A forty-five-year-old contractor is tracked dealing with stressful situations at work, eating a hamburger for lunch, and subsequently throwing ball with some of his colleagues. Every few minutes, the camera zooms in on his chest, and we switch to virtual images from inside his heart and intestines. The endoscopic "camera" takes the viewer on a roller-coaster spin through various organs, and the whirling images of the man's abdomen look as realistic as the images of the man throwing a ball. A commentator's voice confirms what our eyes see: the main arteries are so clogged that the heart can no longer keep the blood pumping. Finally, we witness the pulse's arrest from inside the heart.

Continual switching between exterior and interior views smoothly equates the virtual and real video images. Unlike video endoscopy, virtual reanimations can take on any plane or perspective, appearing to fly through walls of organs. As Steve Connor argues, in the epoch of electronic media, the actual skin is dissolved away and replaced by a "polymorphous, infinitely mobile and extensible skin of secondary simulations."[30] In other words, immersion in a virtual body seduces viewers into adopting a sense of proprioception that they can physiologically never adopt. As virtual endoscopy dissolves the natural borders between organs, it consequently erases the distinction between the real, the mediated, and the "fantastic" (or virtual) body. Although immensely seductive, the notion of intensified immersion or experience through virtual reality is misleading. To make the fly-through feel real, we have to use our imaginative capacities. We can never verify these digital reconstructions within an experiential frame of reference, because no physical or electronic eye can actually travel through organs. The scenes enacted in *Body Story* are obviously fictive; but are the virtual simulations (even though derived from "real" bodies) enough to warrant the label "documentary"?

Docudramas like *Body Story* enhance the illusion that virtual body inspections are painless and traceless; more than that, medical diagnostics and interventions appear to be pleasurable and even adventurous. It should come as no surprise, then, to find that corporeal fly-throughs are modeled not just after "real" video endoscopies, even though they still look like them. Televised virtual endoscopies are also stylized to fit the format of popular attractions in the entertainment industry. The roller-coaster flight down the esophagus would not work if we did not have the experience of Disney World's Space Mountain or Body Wars.[31] Virtual endoscopies perhaps give more insight into our representational conventions than into the "real" body. They are, as Finnish media theorist Erkki Huhtamo calls them, "technological-metapsychological machineries for producing certain cognitive and emotional states of mind."[32] Both as a medical diagnostic and as a televised product, virtual endoscopy is a technological expression of a cultural desire. Computer-assisted gastroscopies signify the apex of ultimate body immersion. Viewers who have grown up in the computer age, as Jay Bolter and Richard Grusin (authors of *Remediation*) argue, are no longer satisfied with representation.[33] They want to be sucked in, soaked up by the object; they don't want to see the organ or molecule, they want to experience what it is like to be inside it. The desire to get past the physiological borders of the body, the organ, and the molecule is synonymous with the desire to surpass the limits of representation. Some would claim that three-dimensional reconstructions stimulate the cross-fading of perception and imagination; sight-seeing turns into site-seeing, as the distinctions between the real, the represented, and the imaginative body fade in the landscape of digital space.

THE ENDOSCOPIC GAZE AND THE UNTAINTED BODY

The history of the endoscopic gaze, as we may conclude from exploring its past, present, and future, should not be viewed exclusively as the progressive development of a medical instrument. The evolution of this gaze is tightly interwoven with the instruments that helped disseminate it in popular culture. In constructivist theory, it is a truism that the history of medical technologies is always intertwined with the history of medical practices; it is equally commonplace to study the history of media technologies—from photography to the computer—in close connection to emerging representational practices. Combining these two fields may seem like a forced coalition, but the emergence of the endoscopic gaze is actually the result of such technological and cultural coevolution. Both developments are anchored in a long tradition in Western culture, and nourished by a belief in enlightenment through technological progress. The movie *Fantastic Voyage* embodies the desire to see, expose, and reconstruct everything subcutaneous while leaving the outer body "untouched." How visualizing technologies changed our view of corporeality may be the lesser-known subtext to the various heroic narratives of doctors as explorers and patients as benefiting subjects. The teleology of more advanced instruments leading to better treatment is contingent on the myths of ultimate transparency and nonintervention. Between the practice of open surgery, aided by primitive gastroscopes, and virtual endoscopic surgery, relying on a variety of digital scanning techniques, we have nursed the fantasy of the "untainted body"—untainted by the instruments of surgeons, and untainted by frequent public exposure to a mass audience.

The epistemological, psychological, and imaginative seduction of the eye steering the viewer's perspective into the endoscopic gaze is rooted in the convergence of medical and media technologies. Epistemological seduction can be traced in the various shifts of the clinical endoscopic gaze. Primitive endoscopes allowed surgeons to look *from the outside in*—the body was accessible exclusively to the surgeon's eye. If she decided to operate, the body had to be cut open. Attaching a camera to the end of a tube brought an end to the surgeon's ocular privilege, as the gaze *from the inside out* could now be shared by patients and others. Video endoscopy reduced the size of the incision needed for the operating instrument to enter the body; seeing and cutting happened through a few tiny holes in the skin. Three-dimensional virtual endoscopy opened up the endoscopic gaze even further: we are now looking *from the inside in,* even permeating the borders of organs. Virtual reconstructions are both representations and projections, and the epistemological distinction between cutting and seeing becomes fuzzy as both acts take place via the computer mouse.[34] To the digital eyes and knife of the surgeon, the body is endlessly accessible and infinitely pliable. Consequently, the viewer is seduced into disregarding the boundaries between interior and exterior, between looking and cutting, and between the real and the virtual body.

The psychological seduction of endoscopic representations can primarily be witnessed in their incorporation by media technologies. More than the X ray or ultrasound, for instance, the endoscope tricks us into believing that we get a perfectly mechanical reproduction of our bodily interior—endoscopy being the landscape photography of medicine. We have now learned, though, that every new visualizing technology raises new problems of perspicuity and interpretation. If anything, the endoscopic gaze has become more complex and opaque as a result of the combination of various digital techniques, each requiring distinct interpretative skills. Whereas, on the one hand, endoscopic technologies give patients more visual access to, and arguably more power over, their own bodies, on the other hand, that same transparency brings with it the liability of having too much access. Video endoscopy changed previous notions of bodily integrity. When the surgeon still had to open up the body in order to see what she had to cut, the body was a relatively private domain; ironically, now that the outer body remains virtually closed during the act of surgery, the interior body has almost become public property.[35] The intimate self is becoming part of a public experience, mediated by surgical and representational instruments.[36]

When patients were intimidated by images of violated body surfaces and laid-open body parts, they would think twice about undergoing a surgical procedure. Seen on television, endoscopic operations look easy, painless, and sometimes even pleasurable.[37] Visual aesthetics define viewers' standards: something that looks appealing and attractive cannot be painful or dangerous.[38] The medical-endoscopic gaze incorporates and defines viewers' trust and belief in medicine's tools and expertise. Combined medical-media technologies render the body transparent, but that transparency is highly filtered by media conventions and broadcast protocols. But due to media exposure, tolerance for surgical procedures has increased. In everyday culture, the body is increasingly hospitable to surgical interference.

Significantly, the body that is increasingly hospitable to surgery is decreasingly tolerant of the scars left by its instruments. The most persistent element in the body-voyage logic (and arguably its most mythical) is the surgeon's purported ability to magically heal the body without scarring its surface. The convergence of optical and surgical instruments represents the "cutting edge" of technology, yet surgeons aim at avoiding just that—cutting the edge (the skin) of the body. Endoscopically assisted surgery metamorphosed from a highly invasive to a minimally invasive to a potentially non-invasive technique. Bodies without borders are the next frontier. With the advent of virtual endoscopic technologies, the interior body seems less a physical object than a digital environment—a site for seeing and playing.[39] The skin is no longer a substantial barrier for surgical entrance, and if the skin can be bypassed, scars seem avoidable. Expectations of a healthy and modifiable interior rise proportionally to expectations of a mint and untainted exterior. As articulated by science-fiction movies and popular television series such as *Fantastic Voyage, Surgeon's Work,* and *Body Story,* reparative medicine holds a magical promise.

Besides the epistemological and psychological impact of the endoscopic gaze, the imaginative seduction of interior body travel constitutes a major factor in the development of endoscopy. Whoever considers *Fantastic Voyage* merely a curious Hollywood fabrication, typical of the Cold War era, severely underestimates the substantial influence this movie had on the medical-technological mindset. Scientists like Richard Robb and Richard Satava explicitly mention the film as a source of inspiration, and with every new stage in the advancement of endoscopy, Fleischer's movie serves as a frame of reference for people to envision this technology. For instance, when the *New York Times* announced on its science page the invention of a so-called endoscopic pill, it used the following opening lines: "They haven't figured out how to squeeze Raquel Welch into it, but scientists . . . have made a pill-size videocamera that takes a fantastic voyage of its own through the digestive tract and transmitting pictures along the way."[40] The story of the miniaturized space capsule once again structures the projected development of a medical imaging technology. In the near future, scientists expect the camera pill to include operating instruments of molecular proportions that will remedy the problems encountered during its inspecting journey through the digestive tract, without leaving a single noticeable trace on the skin. "Not quite the stuff of Hollywood dreams," the journalist concludes, but the result of converging nanotechnology and medical imaging techniques.

Medical instruments, media technology, and technological imagination are inextricably intertwined in the new endoscopic gaze—the gaze from within that conceptualizes the body as a permeable entity. Multiple new cameras open up private bodies to professionalized medicine, engendering a new medical gaze that conjoins public with intimate. As we saw in chapter 2, in a culture that increasingly concedes grounds to public cameras, medical-ethical issues are concurrently media-ethical concerns; the ethics of representation, therefore, are part and parcel of the aesthetics of showing bodies without borders. The promise of a perfectly painless and stainless body voyage should hence be analyzed concurrently as a medical ambition, a popular cultural production, and a collective desire.

CHAPTER 5

X-RAY VISION IN THOMAS MANN'S *THE MAGIC MOUNTAIN*

"'MADAME KNOWS EVERYTHING; MADAME IS WORSE THAN THE X-RAYS' (SHE PRONOUNCED 'X' WITH AN AFFECTATION OF DIFFICULTY AND WITH A SMILE IN DEPRECATION OF HER, AN UNLETTERED WOMAN'S, DARING TO EMPLOY A SCIENTIFIC TERM) 'THEY BROUGHT HERE FOR MME OCTAVE, WHICH SEE WHAT IS IN YOUR HEART.'"[1] IN MARCEL PROUST'S *SWANN'S WAY,* SET IN THE FRENCH ARISTOCRATIC MILIEU AROUND 1900, THE NARRATOR POIGNANTLY EVOKES THE MAGICAL POWERS ASCRIBED TO THE X-RAY MACHINE. THAT THIS newly introduced technology enabled doctors to look through the skin and detect diseases and broken bones was one thing; that it could actually pierce the body and reveal one's most intimate life—"what is in your heart"—was quite another matter. And yet, Proust's mockery of the penetrating quality of Madame Octave's eyes is not just a metaphorical ploy. For a few decades after Wilhelm Röntgen's discovery of X rays in 1895, the meaning and value of the new technology was not exclusively medical but had a much wider resonance, as the culture at large tried to make sense of its mysterious transparent qualities, adjusting the spectator's gaze to a changing scopic regime prompted by new representational technologies. X ray was just one of a range of new technologies emerging in the late nineteenth century that produced realistic or "exact" representations of bodies.

The discovery of X rays, in hindsight, marks an important moment in the history of medicine. In the wake of positivism, scientists like Claude Bernard and Louis Pasteur promoted objectivity, methodological consistency, and standardized observations as guiding principles in both medical science and practice.[2] The objectification of diagnosis through X rays significantly improved the scientific status of medicine. Many believed that, in order to detect and treat infectious diseases such as tuberculosis, mechanically produced pictures were more objective and thus more reliable than other diagnostic methods.[3] Before the advent of X ray, identification of tuberculosis had predominantly relied on sensorial impressions, such as hearing and touching; the new device purportedly allowed the disconnection of diagnosis from an "embodied" perception of symptoms.[4]

Although today the X-ray machine is commonly found in a clinical environment, its value as a medical technology was not self-evident from the beginning. Röntgen's machine came of age in the context of other representational techniques, such as photography and film. Whereas photography and film rendered documentary images of the exterior body, X-ray technology appeared to render the interior body transparent. It was widely assumed that "skiagraphs," as X-ray photos were called, revealed aspects of the inner body that the naked eye could not perceive, such as intimate feelings or immanent death. Besides photography, X-ray technology emerged in conjunction with other representational apparatuses, such as the movie projector and the gramophone. Each of these apparatuses enabled a "mechanical" inscription of the body's image or sound, the realistic quality of which yielded the illusion of physical presence. During the decades around 1900, the X ray acquired its specific meaning and function in the context of a gradually changing audiovisual culture, in which every new instrument redefined the interrelations between observers (or listeners) and objects of inscription.[5] As Lisa

Cartwright argues, the X ray served as a cultural apparatus that "confounded the distinctions between scientific discourse, high-arts, and popular culture."[6]

The initially ambiguous role of X ray as an instrument of both scientific imaging and artistic representation is strikingly illustrated in Thomas Mann's *The Magic Mountain.*[7] When Mann published his novel in 1924, many critics read it as a satire on medical treatment in contemporary sanatoriums and the life of tubercular patients. The story, covering Hans Castorp's seven-years' stay in the Swiss Alp town of Davos, indeed contains sharp and detailed descriptions of therapeutic methods in the heyday of tuberculosis. I argue, however, that Mann's critique of the preferred diagnosis and treatment of the disease was part of a sophisticated interrogation of a new visual regime in medicine and culture evolving at that time. The status of X ray as an objective tool of science is seriously contested in the novel, yet, at the other end of the spectrum, the author also questions the quasi-scientific power of tools that supposedly help visualize psychic or psychological manifestations of the body. More importantly, Mann systematically describes Röntgen's instrument in relation to other representational devices (thus questioning a simple notion of objectivity and showing that mechanically acquired objectivity is a complex and disputable notion), refuting the then dominant "what you see is what you get" adage. New technologies, in fact, redefine the boundaries and interconnections between body and representation, between object and viewer, and between science and art.

X-RAY PHOTOGRAPHY AS A MEDICAL DIAGNOSTIC

In 1882, Robert Koch isolated the causal agent of tuberculosis, the tubercle bacillus. Although doctors then had identified the culprit of the endemic disease that had been the major cause of death throughout the eighteenth and nineteenth centuries, the problem of its diagnosis, treatment, and prevention was still nowhere near solved. By 1907 (the year Hans Castorp enters the sanatorium in Davos), tuberculosis had been recognized by most European states as a national problem that was remedied by organized prevention campaigns, institutionalized treatment, and coordinated research efforts.[8] The invention of X-ray technology added a new weapon in the battle against consumption (as it was then called); it quickly came to serve as a diagnostic aid that helped identify tuberculosis in the early stages of infection, and assisted in monitoring the disease as it evolved. Fifteen years after its introduction, the new machine was among the standard equipment of sanatoriums.[9] Despite the fact that physicians quickly assumed that X rays had diagnostic value, the acclaimed superiority of X rays over other modes of diagnosing consumption did not remain uncontested.[10]

The most radical change induced by the X ray machine was the replacement of the doctor's subjective sensorial impressions by supposedly objective visual evidence. The contest between visual and nonvisual diagnostic means is extensively illustrated in *The Magic Mountain.* When the twenty-three-year-old Hans Castorp enters the sanatorium Berghof, not as a patient but as the visitor of his afflicted cousin Joachim Ziemssen, he is immediately tagged as a likely sufferer of pulmonary tuberculosis because he looks pale and is quickly agitated. The borderline between health, infection, and disease is gradual and diffuse, so the entire range of diagnostic tools is applied to verify the presence of the invisible bacillus. The laboratory inspects Castorp's sputum for bacteriological confirmation, Castorp starts checking his temperature for signs of infection, and the specialists at Berghof perform auscultation and percussion. These last two methods rely exclusively on the doctor's ability to listen with the stethoscope and to tap the chest. When Ziemssen is tapped by Hofrat Behrens, the head of the clinic, Dr. Krokowski enters his colleague's auditive perceptions in a book, "just like a tailor's assistant" would write down the measurements of a person.[11] Although terms like "faint," "diminished," "vesicular," "rough," and "rhonchi" seem like objective observations, they are in fact subjective descriptions based on relative impressions. Sound and touch are (still) considered valuable procedures to justify medical suspicion; since the discovery of X rays, however, they are no longer considered solid *scientific* proof of a latent disease. After Behrens has tapped Castorp and has diagnosed dullness and bronchial sounds in the lungs, he concludes that "the X-ray and the photographic plate [have] yet to come before we can definitely know the facts."[12] Only the X ray will provide "positive knowledge" of the disease, as it will make the effect of the invisible bacillus manifest. Castorp proudly shows his diapositive, on which Behrens has detected "strands" and "nodules," to everyone in the sanatorium. The X-ray photo is regarded as definite proof of the presence of the bacillus, and Castorp becomes a "certified member" of the Berghof community.

In 1907, admission to a sanatorium was based primarily on such radiographic evidence; private sanatoriums in particular preferred to admit patients in the early stages of the disease—patients whose good prospects for recovery would keep mortality statistics down.[13] Not long after its introduction in sanatoriums, X-ray technology became the gold standard for the diagnosis of tuberculosis. Unlike auscultation and percussion, X rays allowed diagnosis without the patient's physical presence; moreover, more than one doctor could "verify" the interpretation.

Consensual interpretation of the pictures, though, was not a given but the result of a long-term process of continual evaluation. It took two full decades of extensive comparisons between new and established methods of diagnosis to show that X rays revealed the same disease, only earlier and better.[14] As radiology emerged as a profession, it became apparent that the center of subjectivity had gravitated from the listening and palpating doctor to the seeing specialist. Body imaging gradually gained reliability as the quality, methods, and protocols surrounding the technology improved, yet the identification of tubercular spots still relied on a human.[15] The black-and-white shadows on the photographic plate were not straightforward "pictures" of spots on the lungs; shadows and silhouettes required substantial interpretation, and misreadings were common. By 1924, the year in which *The Magic Mountain* was published, radiologists still acknowledged a 33 percent difference in interpretation among specialists in the X-ray-based diagnosis of consumption.[16]

Thomas Mann's novel provocatively foregrounds the inherent ambiguity of X rays by stressing their multiple interpretations by so-called experts. Castorp's Italian friend and fellow Berghof patient Settembrini, who is critical of scientific dogmas and religious doctrines, doubts their validity as irrefutable evidence of the disease. When Castorp proudly shows him his diapositive, Settembrini is very skeptical. Having seen hundreds of such pictures, he thinks the decision of whether they offer definite proof of tuberculosis is "left more or less to the discretion of the person looking at them."[17] Settembrini considers the shadows on the X ray open for interpretation, much like art. In his opinion, an experienced layman could master the art of shadow-interpretation just as well as the doctors at Berghof.

Besides replacing doctors' perceptions, X-ray machines overruled patients' own experiences of the disease's symptoms. A patient's experience was inherently subjective, and hence unreliable. With the advent of X ray, the picture—or, more accurately, the doctor who interpreted it—came to define the difference between illness and health. Several years into his *Liegekur,* the doctors decide that Hans Castorp is cured and thus no longer needs to stay at Berghof. When Castorp refuses to believe them, they try to convince their patient by showing him his new X ray: "Here is your latest photo, take it and hold it up to the light. See there! The sheerest pessimist could not see very much in it to find fault with."[18] His symptoms, Castorp is told, no longer refer to tuberculosis, but only to streptococci. Indignantly, he dismisses the doctors' diagnosis and refuses their interpretation. Now that the diapositive shows there to be nothing wrong with him, he concurs with his Italian friend, who doubted it as incontrovertible evidence in the first place.

We see the power of the objective picture in reverse here: whereas the first X ray made Castorp feel sick, the last picture cannot undo that effect. Rather than viewing the diapositive as visual proof of the absence of disease, he doubts the experts' judgment. In spite of his doctors' new diagnosis, Castorp stubbornly continues his life as a patient at Berghof—regularly checking his temperature and observing obligatory rest hours.

The above scene from *The Magic Mountain* illustrates how medical professionals at the time began to trust and prefer mechanically produced images over subjectively perceived symptoms. Part of the X ray's power rested in its ability to "speak for itself," to offer a representation of the inner body apart from its referent in time and space, thus opening it up for intersubjective perception. In that respect, Röntgen's device resembles a contemporary technological invention: the gramophone.[19] When first introduced to the record player at Berghof, Hans Castorp immediately understands the apparatus's significance. The needle reading the record's groove magically reproduces the sounds of singers whose bodies are absent, but whose voices nevertheless harmoniously fill the room. Although the vocalists reside in faraway places like Vienna and St. Petersburg, Castorp is in the presence of "their better parts, their voices."[20] Precisely because of their physical absence, he is capable of experiencing the music in its "purest form." He soon begins to love the voices-without-bodies as autonomous art, even rejoicing in the technical impurity of music that relies on electrical mediation.[21] Castorp keenly observes that, although a record produces the same sounds every time it is played, its perceived beauty varies from person to person, just as his own appreciation changes over time. His favorite albums become cherished objects that he keeps locked in a case and to which he turns for consolation. While the X-ray machine provided him with a reproduction of his inner chest, the gramophone reproduces the lung capacity of altos and sopranos; in both cases the physical content of the torso is transformed into something visual or audible. Once separated from their signifying bodies, the representations facilitated by these technologies function as scientific or artistic manifestations, the ultimate valuation of which fully resides in the eyes or ears of the beholder.

VISUALIZING PASSION

Comparing X-ray technology and the gramophone may perhaps seem a little far-fetched, yet the meaning of this parallel becomes profound when we consider the analogy between X rays and photography.[22] Röntgen's invention found an eclectic range of popular applications related to photography, particularly the fluoroscope, a commercially exploited machine wielded by street vendors, which showed the bones under one's skin. The very first X-ray photo to be published all over the world—in both medical journals and popular magazines—was a picture of Bertha Röntgen's bony hand featuring the sharp contours of her wedding ring. Soon thereafter, the bony female hand became a fetish object; many women had their hands x-rayed to give to their loved ones as "intimate photographs."[23] It was three decades after their discovery that scientists confirmed the danger of excessive exposure to X rays. After this, most frivolous uses of the technique were abandoned.[24]

It was a popular tenet in the early twentieth century that X rays could show much more than bones in a living body; the cathode rays could purportedly expose a person's naked character and uncover his or her deepest passions. The power of the mechanical eyes penetrating the inner chest was thought to extend to the emotional realm, where it could be used to visualize the secrets of the heart.[25] In addition, X rays were associated with sexual intimacy and the revelation of erotic body parts. Popular magazines published cartoons of women visually stripped of their clothing by voyeuristic men wielding X-ray devices. The X ray, as Lisa Cartwright points out, became a socially transgressive instrument that "exposed the private interior of women to the gaze of medicine and the public at large."[26] In *The Magic Mountain,* this voyeurism adds a homoerotic touch to the relationship between the two young men. When Hans Castorp joins his cousin in the X-ray room for the first time, he is completely embarrassed when the doctor invites him to take a look at Ziemssen's heart. The sight of his cousin's inner chest arouses profound emotion and feelings of indiscretion.[27] As the symbolic locus of love and affection, the heart is the most private organ, something one carefully tries to hide from the prying eyes of others. Castorp's embarrassment reflects the popular contemporary idea of X rays as snapshots of the intimate self.

It is no coincidence that this cultural connotation surfaced in sanatoriums in particular. Tuberculosis, as Susan Sontag argues, was often associated with "diseased love" or a passion that "consumes."[28] Intense romantic passions and intimate relationships ("cousining") between patients in sanatoriums were very common, especially since men and women lived in such close quarters, and were inevitably concerned with the one thing that brought them together to that place.[29] Nowhere were the medical and cultural meanings of X rays more intertwined than in the context of the sanatorium. In *The Magic Mountain,* the close connection between love and disease crystallizes in Hans Castorp's adoration of Clavdia

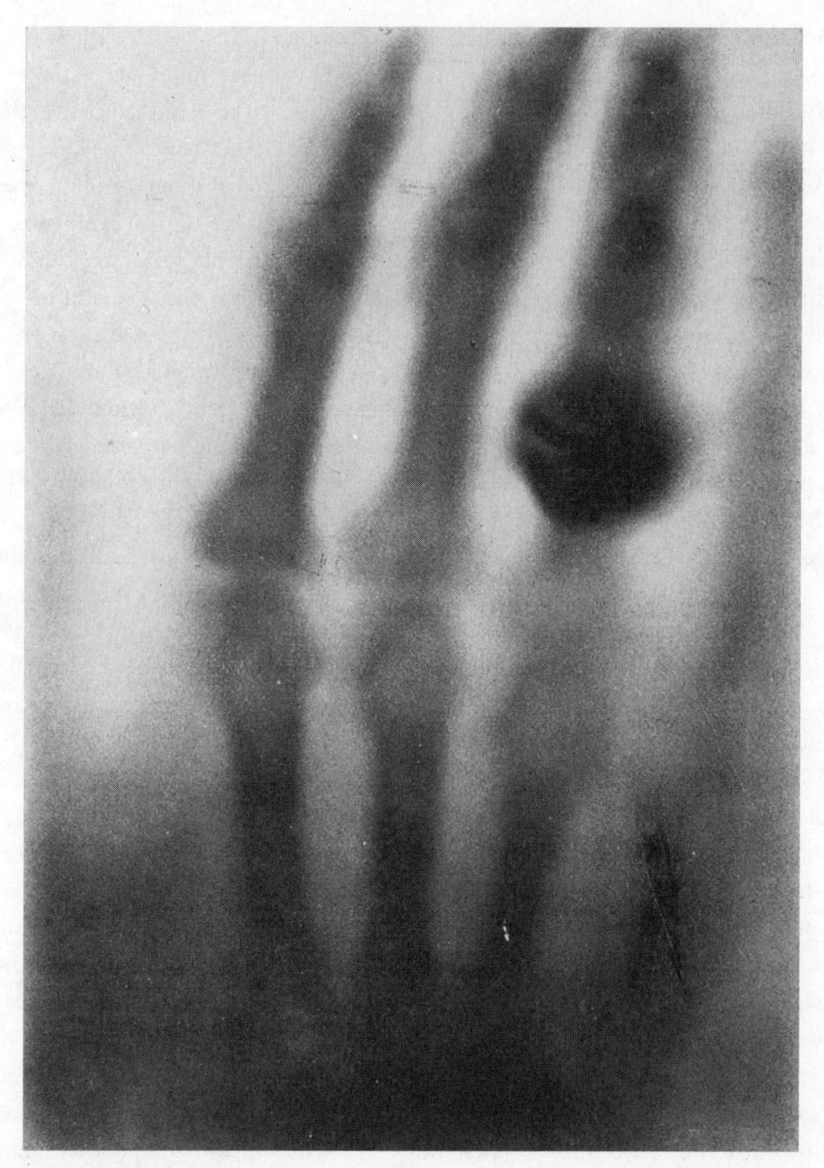

Fig. 15. Bertha Röntgens's hand with ring, produced on Friday November 8, 1895. National Library of Medicine.

Chauchat, another Berghof resident. Every time he thinks of her, his heart beats faster and his temperature rises. Castorp revels in their shared suffering from the same disease, and is convinced that disease is nothing less than transformed love. When he finally engages in a dialogue with the object of his veneration, he declares that Hofrat Behrens "found signs of my love for you in my chest and that he called it 'illness.'"[30] In a long-winded tête-à-tête with his beloved, conducted mostly in French, they talk about the ability of X rays to reveal a person's illness and character. Madame Chauchat carries her diapositive in her wallet, but, like Settembrini, is quite skeptical of its powers to show what is really under the skin. She questions most of all a machine's ability to read the human psyche. Without even looking at Castorp's X ray, she sums up his character and wittingly adds, "Voilà, ta photographie intime, faite sans appareil. Tu la trouve exacte, j'éspère?"[31] At the end of their conversation, Castorp eventually wins the trophy he was after: Clavdia gives him her X ray before leaving the sanatorium for extensive traveling back in her home country. The diapositive obviously means nothing to the woman, but to Hans Castorp it signifies that she has now put her intimate self into his hands; her X-ray photo means more to him than would an ordinary photograph of his beloved.

Clavdia's X ray, as it turns out, is anything but a straightforward inscription of her love, just as Hans's own medical X ray was anything but an unambiguous inscription of his disease. After her departure, Hans carries her shadowy picture in his breast pocket and regularly takes it out, trying to reassure himself of their mutual feelings: "Then he flung himself into his chair, and drew out his keepsake, his treasure, that consisted, this time, not of a few reddish-brown shavings, but a thin glass plate, which must be held toward the light to see anything on it. It was Clavdia's X-ray portrait, showing not her face, but the delicate bony structure of the upper half of her body, and the organs of the thoracic cavity, surrounded by the pale, ghostlike envelope of flesh."[32] During Clavdia's long absence, the X ray becomes a substitute for her body. Upon her return, he finds out, to his great distress, that the diapositive meant different things to each of them. In an emotional conversation, Castorp admits his love for her. Clavdia mocks his worship of her X-ray portrait, dismisses the fetish object, and in a dramatic gesture kisses him on the lips, only to tell him that she does not love him. She convinces Castorp that the word "love" has several meanings—spiritual, physical, religious—which vary depending upon the interpreting subject. When the doctors told him there was no longer any evidence of tuberculosis in his chest, Castorp refused to believe them; he now refuses to believe that his love for Clavdia may have never been "real."

Even though the X ray was never a visual imprint of her inner self, nor of her love, the object had a real impact on Hans Castorp's body. "For love of her . . . I declared myself for the principle of unreason, the *spirituel* principle of disease, under whose aegis I had already, in reality, stood for a long time back."[33] X rays made both illness and love manifest, and both corporal manifestations filled him with spiritual energy. The love, just as the disease of tuberculosis, may have been an illusion, but the effect of the skiagraph cannot be undone.

The contest between X ray and sensory forms of diagnosis as the best representation of the inner body, is reminiscent of a similar contest between painting and photography as the best representation of the outer body.[34] In the mid-nineteenth century, portrait painting was gradually replaced by portrait photography, and a number of artists who had opened portrait photo studios made a fortune. A photograph was considered more lively and true-to-nature than an oil painting on canvas. Along the same lines, an X ray was seen as an enhanced photograph: a portrait that revealed everything under the skin. Among the first technicians to operate the Röntgen apparatus were photographers without any medical training. They opened up "X-ray studios" and called their photo sessions "X-ray sittings." It wasn't until the 1910s that radiology emerged as a medical profession, thus forcing the photographers out of business. Artistic use of X-ray techniques remained marginally acceptable until the true dangers and hazardous effects of radiation were acknowledged. Unregulated X-ray studios were then outlawed, and the darkrooms moved to hospitals and treatment centers where operation of the machines was restricted to medical doctors. Eventually, X ray achieved the status of a scientific imaging technology rather than an artistic one. As science writer Bettyann Holtzmann-Kevles concludes, "with the appearance of radiographs, the ideal of an objective diagnosis became a major element in the transformation of medicine from an art into a science."[35]

The Magic Mountain symbolically enacts this historical friction between artistic and scientific uses of the X ray. When Hans Castorp first enters the dark anteroom of the X-ray facility, Hofrat Behrens introduces his patient to what he calls his "picture gallery": rows of dark plates on a wall, each showing vague and shadowy portions of a human body in which the sharp nucleus of a skeleton stands out. Without further indication, it would be impossible to identify these portraits. Behrens turns out to be both a radiologist and an amateur painter, and Hans Castorp immediately relates these two skills. Berghof's head physician has made two portraits of Castorp's beloved Clavdia: an oil painting and an X ray. After Castorp successfully fishes for an invitation to take a look at Behrens's artwork, which he exhibits in his private quarters, the row of paintings strikes him in the way that the X-ray gallery did. Clavdia's portrait shows only a distant likeness to the real woman; nevertheless, he praises the portrait's realism and precision. Although the painted Clavdia is anything but lifelike, Castorp still feels like he might smell the odor of her body by pressing his lips on the canvas, and wonders how the doctor has managed to capture the essence of his object of passion. Modestly dismissing his patient's praise, Behrens states that the verisimilitude of the portrait is due mainly to his intimate knowledge of her inner body: "I know her under her skin—subcutaneously, you see: blood pressure, tissue tension, lymphatic circulation, all that sort of thing."[36] Castorp finally realizes that the one thing that makes this painting special is the meaning he himself attaches to it. "You paint what isn't really there and yet it is," he tells Behrens.[37] Beauty or radiance is not in the painting but in the eyes of the beholder, just as the identification of health and illness is not in the X ray but in the eyes of the doctor. *The Magic Mountain* explicitly links observer variability in the medical interpretation of X rays to the subjective interpretation of art.

This scene is more than a mere description of the historical wager between X ray, photography, and the painted portrait; Mann's novel passes judgment on the growing gap between "the two cultures," art and science. At a time when the medical use of X-ray photography was steadily gaining ground, the novel questions its supposed objectivity and therefore scientific superiority. The main concern of the study of medicine, as Hans Castorp concludes, is also the concern of the arts: the human being. Artistic and medical gazes have always been closely associated, to the extent that even Greek sculptures are grounded in profound knowledge of the human body. "You can see how the things of the mind and the love of beauty come together, and that they always really have been one and the same—science and art."[38] The lyrical, technological, and medical are all variations of the same pressing human concern, which Castorp regards as schools of humanistic thought without assuming any hierarchy among them.

THE PHOTOGRAPHY OF DEATH

Perhaps more than penetrating the flesh to reveal agents of disease, and more than baring the secrets of the heart, Röntgen's new device allowed people to steal a glance at their future fate as a skeleton. The shadows of bones on skiagraphs were strongly associated with mortality; death was imprinted in the living body and X rays made it visible to the naked eye. In 1896, medical journals began publishing composite skiagraphs of entire human bodies, adult and child, even if these skeleton pictures served no scientific purposes whatsoever. The new rays were not only credited with penetrating powers—they rendered the flesh transparent—but they were also thought to have predictive qualities: literally foreshadowing man's deadly destination and turning the body into a transcendent object.

When Castorp first glances at the scaffolding of his own mortal flesh, by means of X rays, he immediately associates it with death. Observing his bony tenement, the young protagonist realizes that he is looking at his own memento mori:

> Castorp saw precisely what he must have expected, but what is hardly permitted man to see, and what he had never thought would be vouchsafed for him to see: he looked into his own grave. The process of decay was forestalled by the powers of the light ray, the flesh in which he walked disintegrated, annihilated, dissolved in vacant mist, and there within it was the finely turned skeleton of his own hand, the seal ring he had inherited from his grandfather hanging loose and black on the joint of his ring finger—a hard, material object, with which man adorns the body that is fated to melt away beneath it, when it passes on to another flesh that can wear it for a little while. With the eyes of his Tienappel ancestress, penetrating, prophetic eyes, he gazed at the familiar part of his own body, and for the first time in his life he understood that he would die.[39]

There is no medical need for Castorp to have his hand x-rayed; the hand, rather than the lungs, is the prime object of visual examination for entirely symbolic reasons. The transience of the flesh, in Castorp's philosophical reflection, is offset by the durability of the golden ring—the only artifact that will survive generation after generation. The x-rayed hand must have reminded contemporary readers of the famous skiagraph of Bertha Röntgen's fingers, on which her wedding ring stood out as the symbol of marriage, faith, and especially the fruitful reproduction of descendants. Death and the continuity of life convolute in the sign of the ringed x-rayed hand.

In the popular imagination, the skiagraph was considered an omen of immanent death—a belief that could be readily dismissed as contemporary superstition. The significance of this belief becomes apparent, though, if we consider how scientists used Röntgen's machine to verify the existence of ghosts and spirits. In the second half of the nineteenth century, as historian Alan Grove explains, the Society for Psychical Research had been established, which aspired to back up the belief in ghosts with sound empirical evidence; through careful scientific investigation, its members hoped to prove that physical phenomena were much more than a popular fantasy.[40] Earlier in the nineteenth century, the photo camera had been the favorite verification tool. It appeared to have the spooky ability to fix things on the plate that were not really there, showing shadows of figures that had not been present when the photo was taken, and, the other way around, sometimes people or things that had been visible on the photographic plate slowly disappeared from the picture. It took a while for people to learn that the ghostly portraits and vanishing figures where nothing but the result of double exposures and fading chemical fixtures. Although they had hoped the photo camera would prove a reliable instrument for the detection of ghosts and spirits, the pictures often disclosed nothing but the technical limitations of the "old photography."

The invention of the "new photography," as X ray was frequently called, instilled high hopes in members of the society. In popular culture, Röntgen's skiagraphs were already frequently connected with eerie images of skeletons. As Grove argues, to accept the X ray as an instrument that could verify the existence of spirits was the next logical step: "Röntgen's discovery immediately confirmed preexisting theories about invisible vibrating forces and, within the popular imagination, seemed to confirm scientifically the existence of mesmeric and ghostly forces in our world."[41] Pairing a firm belief in supernatural phenomena and spiritualism with an equally strong scientific rationalism was not a contradiction at that time. Quite the opposite, imaging technologies were hailed as necessary to help prove beyond reasonable doubt that spirits and ghosts indeed existed as immaterial substances in our material world.

The Magic Mountain explicitly deals with the tension between spiritualism and science, most prominently in the chapter titled "Highly Questionable" near the end the novel. Various critics have dismissed this chapter as frivolous, or considered it incompatible with the serious and realistic content of the rest of the novel.[42] Others have related Mann's temporary interest in the occult to the momentum of German nationalism's infatuation with national myths and superstitious narratives.[43] In my view, however, the chapter perfectly fits with Mann's thematic interrogation of new technologies as cultural apparatuses that affect dominant scopic regimes. Whereas Dr. Behrens embodies the continuity between the professions of painting and radiology, Berghof's other attending physician, Dr. Krokowski, personifies the (in our eyes compromising) alliance between science and spiritualism. A "modern" scientist who is fascinated by Freudian psychoanalysis, Krokowski extends his studies to the more questionable realms of parapsychology and the occult. Although some Berghof residents frown upon Krokowski's resort to mysticism, he defends his preoccupation with the regions of the human soul as a purely scientific interest in the subconscious. Intrigued by the relationship between spirit and matter, Krokowski articulates his belief that all organic matter must be seen as some organizational form of energy, and it is precisely this existence of ethereal substances and energy flows that he wants to prove scientifically.

In line with the practices of the Society for Psychical Research, Krokowski considers the X ray as the tool par excellence for tracing the transformation of the flesh. He conducts his investigations into parapsychological phenomena with the rigor and thoroughness of an experimental physicist, observing methodological protocols and relying on the objectivity of mechanically produced pictures. His quest for scientific validation begins with the arrival at Berghof of a supernaturally gifted patient, Ellen Brand. The nineteen-year-old Danish girl claims to have supernatural abilities: she can look through walls and people's clothing, and can see objects and hear people's conversations from another room. In addition, she has "visions, both visible and invisible," that allow her to "see" her sister's death even though they live on different continents.[44] Brand claims she is a medium for a spirit named Holger, who brings her into contact with the world of the dead. Krokowski extensively examines the girl and subjects her to various X-ray sessions—and not just to test her for tuberculosis.

At the end of the novel, Hans Castorp has become as skeptical about X ray's ability to verify ghosts as about its ability to prove the presence of tuberculosis. Unlike Dr. Krokowski, Castorp does not take the authoritative power of the X ray for granted, but assumes the role of watchdog. At a séance, attended by both Castorp and Brand, inexplicable forces are set in motion: someone gets slapped in the face and suddenly the light is turned off. But the strangest thing happens to the protagonist when something falls into his lap: "He discovered it to be the 'souvenir' which had once so surprised his uncle when he lifted it from his nephew's table: the glass diapositive of Clavdia Chauchat's X-ray portrait. Quite uncontestably, he, Hans Castorp, had not carried it into the room."[45] Reluctant to accept the X ray as a scientific tool, he keeps quiet about the incident. When he later discusses his experience with Settembrini, his friend dismisses Brand as an impostor and the ghost as a delusion, warning Hans that an engineer like himself should only rely on "his power of clear thought."[46]

The Magic Mountain critically evaluates the peculiar coalition between a positivist belief grounded in the authority of the instrument and a romantic conviction cemented in subjective impressions. Dr. Krokowski represents the former, while the residents of Berghof adhere to the latter position. Castorp seems equally suspicious of both views, and, more than anything, he distrusts the alliance between an unconditional faith in scientific method and an uncritical acceptance of intuition. Despite Settembrini's warning to stay far from such dubious enterprises, Castorp decides to attend a séance in which Ellen Brand, instigated by the head physician, tries to invoke the spirit of his deceased cousin. The dimly lit séance room reminds Castorp of the X-ray lab, where he had once asked Ziemssen permission to commit the "optical indiscretion" of looking into his heart: "He liked the darkness, it mitigated the queerness of the situation. And in its justification he recalled the darkness of the X-ray room, and how they had collected themselves, and 'washed their eyes' in it, before they 'saw.'"[47] An hour or so into the séance, Ziemssen's ghost appears as a shadowy figure in military uniform with "a warrior's beard and full, curling lips."[48] Instead of speaking to Ziemssen, as Krokowski urges him to, Castorp stands up and turns on the light, and the ghost promptly disappears before Krokowski can take a photograph or X ray. The physician, of course, is angry that he has spoiled the experiment, yet Castorp is convinced that no picture can prove the existence of his deceased cousin. The visualization of the spirit, like the materialization of tuberculosis and inner feelings in an X ray, is "highly questionable" in various meanings of the word. Besides the fact that he considers taking a picture of the dead very inappropriate, he also strongly doubts the photo's status as objective proof of an imperceptible reality. Castorp contemptuously dismisses the attempt to photograph his dead cousin, because he wants to hold on to his memory of the living Ziemssen, as he wanted to hold on to the former diagnosis of his disease and his love for Clavdia Chauchat—even though X rays also in these cases delivered "incontrovertible evidence" to the contrary.

Thomas Mann, in the words of his protagonist, rejects the spiritual invocation of ghosts both as irrational, romantic aberrations and as rational, scientific experiments. In an essay, Mann reported his own experiences with séances in Munich, expressing his ethical reservations.[49] In *The Magic Mountain,* the author articulates his skepticism in a symbolic fashion. Mann provides a striking description of X ray's ambiguous denotations and powerful cultural connotations at that time, and the novel's open ending seems to underscore this preference for ambiguity rather than closure.[50] In the next and last scene of the book, Castorp leaves the sanatorium, joins the army, and ends up in the trenches of the battlefield. Previous looks into the bony X-rayed chest of the protagonist, combined with the ghostly appearance of his cousin Joachim during the séance, can be read as an overture to Castorp's own death. However, the narrator never states this explicitly, he just "leaves him there"—leaving his fate to the reader's imagination.

THE NEW MECHANICAL GAZE

In the first two decades of the twentieth century, the novel's time frame, the X-ray machine had not yet acquired its distinctive medical function; its cultural connotations and uses were still ubiquitous and resonant in popular culture. The dissemination of X-ray images was infused with a wide range of scientific, social, and cultural inscriptions. People endowed the machine with the capacity to produce objective visual inscriptions of imperceptible manifestations of the inner body, and, by the same token, they believed that the penetrative eye of the new technology would reveal the most intimate secrets of the human soul. The new technology was just one in a long series of devices that separated the actual body—both in time and space—from its representation. In *The Magic Mountain,* X-ray machines are subtly connected to other representational technologies that were still new at the time and that still conveyed an aura of magic: photography, the gramophone, and even the cinema capture the imagination of Berghof patients.[51] As German media theorist Friedrich Kittler has so powerfully argued, the emergence of these technologies should be studied conjunctively in order to understand historical issues of mechanical reproduction and contemporary notions of objectivity.

X-ray technology, like painting and photography, is a representational technology creating an illusion of unmediated, objective reality. During his seven-year stay at the sanatorium in Davos, Hans Castorp (and in his wake, the reader) gradually comes to realize that the scientific and technological innovations of the modern era do not live up to their purported objectivity. As with photography, X rays enable the viewer to see more, but mold as well as reflect visual reality. As a consequence, Castorp refuses to accept the scientific paradigm that claims primacy of vision as the establishing principle of truth. What is real or true largely depends on the interpretation of the observer, and the highest attainable goal is to reach a consensus about what a representation stands for. An increasingly dominant positivist belief system assigns absolute superiority to scientific experiments and objective instruments, but Mann shows that in such a system, perception merely shifts from the subjective observer of the body to the intersubjective observer of mechanically induced representations of the body.

The hero of *The Magic Mountain* develops a profound skepticism towards the all-encompassing claims of positivist science, but also towards the irrational and the occult. This criticism manifests itself in the way Mann describes the technical instruments of both science and art. As representation technologies, they are part of an audiovisual culture that, at the turn of the century, rapidly mechanized the gaze. Rather than endorsing a hierarchy of rationalist, objective scientific instruments and emotional, subjective, artistic technologies, the author emphasizes their diffused continuity. The body—made transparent by a host of new mechanical instruments—is anything but an objective object of study. On the contrary, X-ray pictures, like other mechanical reproductions, always yield mediated perspectives, as their meanings are always shaped by the knowledge and feelings of their interpreters. The X ray, as the "enhanced eye" of art and science, is a small but significant element in the emergence of a new way of seeing, a visual regime that believes in transparency and translucency. As art historian Jonathan Crary has suggested, new technologies were part and parcel of a cultural reconfiguration of relations between observing subject and modes of representation, a transformation he refers to as "modernizing vision."[52] Thomas Mann's profound interrogation of the supposed objectivity of X rays in the context of other representational technologies forces the reader to reconsider the assumed dichotomies between instrument and observer, between object and representation, and between science and art.

CHAPTER 6

ULTRASOUND AND THE VISIBLE FETUS

AN EPISODE OF THE POPULAR TELEVISION SERIES *ER* ENDS WITH A REVEALING SCENE. PETER BENTON, AN AMBITIOUS SURGICAL RESIDENT, HAS VIRTUALLY IGNORED HIS PREGNANT GIRLFRIEND, CARLA, WHEN SHE IS ADMITTED TO THE ER BECAUSE OF EARLY CONTRACTIONS. AFTER THE GYNECOLOGIST ON DUTY HAS FINISHED CARLA'S ULTRASOUND EXAM, THE PATIENT IS FAST ASLEEP WHEN PETER, STILL WEARING HIS BLUE SURGEON'S GOWN, QUIETLY SLIPS INTO THE ROOM. HE CASTS A QUICK GLANCE AT HIS GIRLFRIEND, YET his attention is immediately caught by the ultrasound monitor still featuring the frozen image of the fetus. With the eyes of an expert, he checks the gray-white shadows and the numbers in the margins of the screen. Then a remarkable metamorphosis takes place: the usually tough Dr. Benton turns emotional as he starts touching and caressing the shadows on the screen. Seeing the sonogram radically changes Peter's attitude towards Carla's pregnancy; from now on, he shows himself passionately responsible for the fetus, which is born prematurely in the next episode.

This dramatization of an everyday occurrence—a future father seeing the first ultrasound of the fetus—illustrates a commonsense fallacy: the clinical gaze of a medical professional precedes the emotional bonding of a parent to its child. Many parents can remember a clinical expert giving the green light before allowing themselves to "recognize their baby" on the ultrasound. In case red flags go up during the prenatal ultrasound, prospective parents wait for the clinical results to inform their (emotional) decision to continue or terminate the pregnancy. A sonogram, used in a clinical context, is obviously inscribed with a variety of meanings—medical, psychological, emotional, and cultural. In series like *ER*, medical meanings can apparently be separated from "other" meanings: Peter Benton first looks at the sonogram as a detached medical professional, and subsequently surrenders himself to its emotional appeal. Because doctor and prospective father converge in the same person, his changing facial expressions demonstrate the "natural order" between the clinical look and emotional attachment.

To what extent can the clinical and emotional meanings of ultrasound be separated? In this chapter, I will describe a case study concerning the past implementation and recent reconsideration of ultrasound in the Dutch prenatal care trajectory from conception to delivery. Between 1985 and the year 2000, the psychological and cultural meanings of this imaging technology gradually merged with strictly medical ones. At the beginning of the new millennium, the ultrasound exam of a pregnant woman is concurrently a medical diagnostic checkup, a psychosocial event, and a photographic ritual. In recent years, OB-GYN specialists in the Netherlands have expressed their dissatisfaction with the gradually "convoluted" status of the first-trimester prenatal scan, and have urged hospitals to restrict the use of ultrasound to "medical purposes."[1] In the following analysis of this recent debate, I contest the view that ultrasound can be restored to its original medical import and purged of its cumbersome nonmedical (emotional, cultural) connotations. Ultrasound is in and of itself an ambiguous technology, and in past decades medical practitioners have actively contributed to its becoming an intricate technical, medical, and sociocultural phenomenon.

At the core of my criticism is the assumption, implicit in the emerging "purification" view, that through ultrasound, pregnant women and their partners obtain neutral or technical information on the future health status of their fetus. Receiving "pure clinical information" supposedly facilitates and rationalizes their emotional decision-making process. The autonomy of the decision maker is thus a corollary to the medical purification of the ultrasound scan. I argue that in the organized everyday practice of ultrasound, technology, doctors, and patients are all constitutive actors in the processes of visualization and signification.[2] The teleology of a sonogram "revealing" information about the fetus, medical specialists "translating" the visual data into options, and the pregnant couple subsequently "making a rational, informed decision" renders a simplified view of a profoundly complex practice—a practice that has evolved historically and cannot simply be undone.

A JANUS-FACED TECHNOLOGY

The implementation of ultrasound as a common diagnostic tool in prenatal care has had anything but a "natural trajectory"; the career of this imaging technology is remarkably rife with contingencies, and is also distinctly shaped by social, professional, and cultural forces.[3] In the public mind, ultrasound is commonly associated with pregnancy, notwithstanding the fact that the technology originated in a completely different field and is still widely used outside the domains of obstetrics and gynecology. Primitive forms of ultrasound emerged in the early twentieth century, when British engineers tried to prevent another *Titanic* disaster by developing a technique to detect submerged icebergs with sound reflection.[4] "Sonar," as this technology was called, also allowed allied forces to spot enemy submarines during World War I. The first experiments with medical-diagnostic applications took place in the 1930s.[5] But the development of ultrasound really took off in the 1950s, after Scottish medical scientist Ian Donald coincidentally discovered the potential of sonar to visualize a fetus at an early stage of pregnancy, and started to apply it to the field of obstetrics.[6] Until gynecologist Alice Stewart, in 1956, proved the harmful effects of radiation on fetal development, X rays had commonly been used to diagnose multiple fetuses up until the thirty-fifth week of gestation. The subsequent ban on X rays of pregnant women accelerated the advancement of ultrasound as a diagnostic device. In 1962, the first ultrasound machines entered the marketplace, and a decade later, virtually all gynecologists in Europe and the United States had equipped their consultation rooms with the new machine.[7] Consequently, the power over the visualization of the fetus shifted from radiologists to gynecologists. Further development of ultrasound technology was the result of a unique collaboration—and also a fierce competition—between medical specialists (gynecologists, radiologists, and surgeons) and engineers.[8]

Ultrasound's major revelation was to show a living fetus in utero; even more significant was the invention that would capture that image on a screen and record it on tape. As early as 1963, Ian Donald understood that the content of the uterus can be better visualized if the pregnant woman's bladder was filled with fluid. A transducer moved across the abdomen sends out sound waves, which are conducted by fluid until they collide with solid organic material like bone tissue. Bounced-back sound waves are then transformed into images—reverberations inscribed in black-and-white shadows on the screen.[9] As soon as the embryo starts to develop bone tissue, the sonar meets resistance, which materializes into white and gray lines on the screen. The first recorded ultrasound images in the 1970s showed little more than white and gray lines—nothing even remotely resembling a baby. The invention of the scan converter in the late 1970s translated the abstract white and gray lines into a more coherent picture, but it was not until video technology came of age that the fetus appeared as a recognizable entity. Real-time video ultrasound was originally invented to save records of ambiguous gray shades for long-term comparative research. The direct screening of real-time images during a prenatal exam, however, had a tremendous yet unexpected "side effect": women loved seeing images of their fetuses moving around in the uterus, on the screen, even if they were quite incapable of interpreting them correctly.[10] After 1985, ultrasound quickly gained popularity as a diagnostic tool, not only because it allowed for a painless prenatal checkup but also because it rendered the first visible proof of the invisible fetus.[11]

It is instructive to linger over the question of what exactly an ultrasound image can show and at what stage of pregnancy. Before the eighth week of gestation, an ultrasound scan yields little more than a heartbeat, because bone tissue is still in its formative stage. Between weeks nine and twelve, a sonogram may reveal ectopic and multiple pregnancies, and show whether the fetus has all putative limbs in the right place. Forty years of ultrasound checkups have yielded average growth curves, so the first fetal biometric exam permits calculation of the delivery date and observation of a possible growth lag. During the second trimester of a pregnancy (weeks twelve to twenty-eight), ultrasound is primarily deployed to detect fetal abnormalities, ranging from serious congenital defects such as spina bifida, anencephaly, or hydrocephalus, to cosmetic irregularities such as a harelip, a cleft palate, or clubfeet. With the help of advanced ultrasound technology doctors can also detect heart and kidney failure.

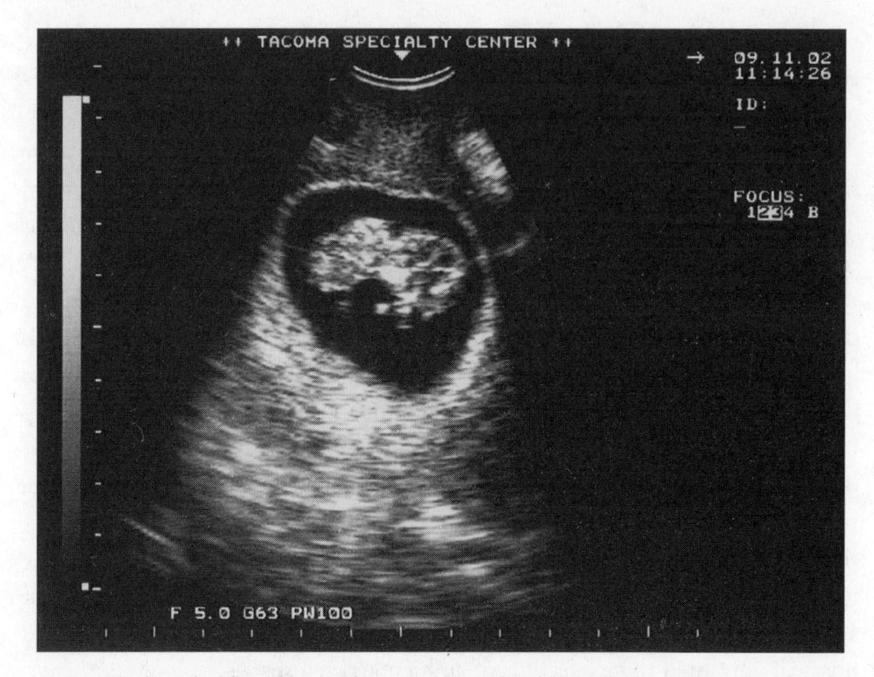

Fig. 16. Ultrasound of first-trimester fetus. Courtesy of Kathleen Pike Jones.

Yet what a sonogram may disclose at each stage of pregnancy is also highly dependent on the technical quality of the instrument. Not every ultrasound apparatus yields the same results; the optical quality of a machine is commonly commensurate with its price.[12] The cheapest machines produce acceptable images for measuring fetal growth, whereas technically advanced and very expensive machines display the subtle shades of gray that allow sonographers to perceive heart or kidney failure. As important as the question of what a sonogram can reveal is the question of what it cannot reveal. Even the best instruments cannot show everything; at best, ultrasound equipment reveals 65 percent of all potential aberrations, which means that 35 percent of fetal defects remain invisible on a scan.[13] Naturally, every technological innovation changes the state of the art, increasing doctors' ability to detect fetal abnormalities at an earlier stage, but the expectation that ultrasound will at once visually reveal each and every defective condition at the earliest prenatal stage is illusory.

Visualization processes in medical practice are never straightforward decoding activities, but are what sociologists of science K. Ammann and K. Knorr Cetina have dubbed "the interactive machinery of seeing."[14] When it comes to reading

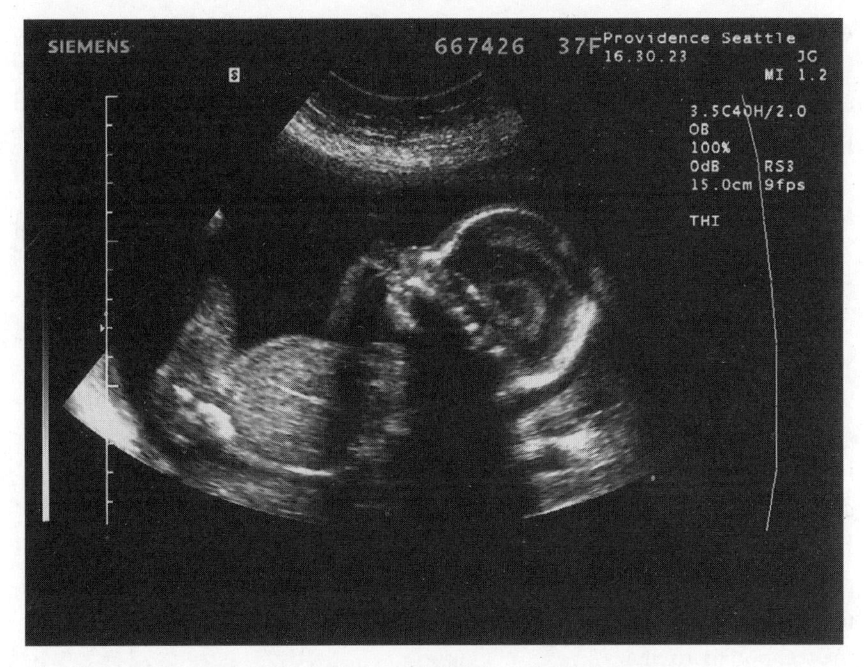

Fig. 17. Ultrasound of second-trimester fetus. Courtesy of Marcelle Garrard.

and interpreting an ultrasound scan, there are distinct levels of sophistication, and the outcome depends on who reads what. Comprehending quantitative results related to fetal growth curves is a relatively simple skill that does not require much training. But in order to detect anomalies and congenital defects on an ultrasound scan, one has to have a thorough knowledge of anatomy and substantial experience reading scans—skills usually possessed only by specialized sonographers. Interpreting a scan is not the same as decoding it: a sonogram shows shades that may or may not be significant. Information on a sonogram is often ambiguous; deciding which information is relevant constitutes the specialist's most difficult interpretive task.[15] It has taken years of negotiation between doctors and technicians, years of readjusting machines and recalibrating imaging protocols, for consensus to take shape.[16] Even so, the sonogram is never an unequivocal predictor. In many instances, the gynecologist is bound to give only an indication of potential complications, a percentage of possible abnormality. Ultrasound does not guarantee certainty; innovations in technology engender new liabilities, new uncertainties, and thus new anxieties.

Ultrasound is also an inherently ambiguous technique because it can generate multiple types of pictures. Moving the transducer across the abdomen enables sonographers to scan the surface and size of the fetal body; scanning the contours and surfaces results in sonograms of figurative pictorial quality. Sonographers who perform ultrasound exams to detect fetal abnormalities primarily look at cross sections of the abdomen rather than surface scans. Without sufficient anatomical knowledge, it is impossible to translate two-dimensional cross sections on the screen into three-dimensional anatomical concepts. Even though both types of scans refer to the same entity, they connote very different mental concepts. Whereas the first type of scan, producing realistic contours, refers to a "baby," the second type of scan elicits mental images of a "fetus"—a product of scientific rationality. Sociologist Lorna Weir has noted that both terms are used in the prenatal clinical setting, but their rules of circulation differ, "with 'fetus' located in co-practitioner interactions, and 'baby' in a line of speech leading from the clinic into the discourse practices of pregnant women and mothers outside the clinic setting."[17] Hence, the ambiguity of the fetus as a rational, scientific object and an emotional part of a subjective self is both constructed and enhanced by ultrasound technology. Ambiguity is not the result of ultrasound's use but is already inscribed in the technology itself.

ULTRASOUND'S DOUBLE EFFECTS

The ambiguity of the ultrasound exam definitively shows in how it impacts the experience of pregnancy. Before the invention of ultrasound, pregnancy was primarily considered an individual experience. Only the pregnant woman could feel the fetus moving; outsiders could only hear its heartbeat.[18] Feeling and listening are still important sensorial perceptions, but ever since ultrasound has entered the prenatal trajectory, sight has arguably become the privileged sense perception. For one thing, ultrasound has opened up the womb—and thus the intimate perceptions of pregnancy—to others besides the pregnant woman herself. Feminist philosophers and anthropologists have extensively theorized about the cultural impact of visualization techniques such as ultrasound on the configuration of the female reproductive body.[19] The sonogram depicts a visible fetus that, once severed from its "maternal environment," allegedly gains autonomy at the expense of the pregnant woman as subject.[20] Ultrasound thus effectively dethrones embodied female experience and "de-genders" the perception of the fetus. Besides theorizing the effects of ultrasound, a number of sociologists have empirically studied the immediate impact of prenatal scans on a woman's experience of pregnancy, but the conclusions, which I will discuss below, are remarkably contradictory and often disputed.

Since the massive implementation of prenatal ultrasound scanning, researchers have tried to prove its beneficial psychological effects: sonograms reassure prospective parents, promote parental responsibility, and facilitate prenatal bonding. Psychologist Beverly Hyde found the most obvious positive psychological effect of ultrasound screening to be parental reassurance.[21] If the ultrasound scan shows a fetus to be normal, parents can relax and enjoy the rest of their pregnancy; if the sonographer detects fetal defects, the woman and her partner can decide to terminate the pregnancy. In both instances, ultrasound is said to offer the parents peace of mind and hence to be a desirable practice.[22] Follow-up measures indicate, however, that reassurance is transient. While some researchers contend that clinical information about the health status of a fetus reduces emotional stress during the pregnancy, others have proved ultrasound scans to increase psychological pressure on prospective parents, who are often confronted with uncertain prognoses.[23] A famous early study, made in 1982, stated that seeing the fetus improved maternal responsibility: pregnant women would drink and smoke less after seeing visible evidence of their fetus, and generally feel more responsible for its well-being.[24] Yet, since the routine implementation of ultrasound in prenatal care, there has been no significant decrease in the use of harmful stimulants by pregnant women. Finally, researchers tried to establish a causal relationship between ultrasound scans and prenatal maternal bonding; their research assumption was based on an earlier study that demonstrated the positive effects of intimate contact between mother and child immediately after birth.[25] Recent investigations have countered these results: increased maternal bonding is less an effect of a woman seeing the fetus in utero than of a woman's positive attitude towards her pregnancy in general.[26] Moreover, maternal bonding through sonography may have a downside; some experts argue that ultrasound scanning may have a detrimental effect on women's coping with grief after a late-trimester medical abortion or miscarriage.[27] Others have shown ultrasound to have a similar effect on the male partners of women who had spontaneous or induced abortions.[28] When prenatal scans result in terminated pregnancies, the visualized fetus appears to give parents a more distinct feeling of loss.

The problem with many of these studies, I believe, is their a priori distinction between clinical effects (the information obtained from an ultrasound on the health status of the fetus) and psychosocial effects (the way the client handles the information). My point is that psychological or emotional impacts and medical effects are extremely hard to differentiate because they have coalesced beyond distinction. The effects of ultrasound on the experience of pregnancy are indeed so complex because the prenatal test concurrently renders a clinical and an emotional message—messages that are frequently at odds. Seeing a normal fetus may reasssure a pregnant woman and encourage maternal bonding, but if the scan shows a fetal defect, that same effect—bonding—complicates or even compromises the bonding process. American anthropologist Janelle Taylor has called the inherent ambiguity of ultrasound effects the "prenatal paradox": "Pregnancy is constructed more and more as a tentative relationship . . . at the same time and by the same means that pregnancy is also constructed as an absolute and unconditional relationship, and the fetus as a "person" from its earliest stages."[29] Almost all inquiries into the effects of ultrasound scanning confirm the "prenatal paradox": if the baby turns out healthy, the scan enhances positive feelings; if something goes wrong during the pregnancy, it augments the negative effects.[30] The interweaving of various effects is not the exclusive result of the ambiguous technology, but also of the way in which that technology is implemented in, and regulated by, national and local policies.

AMBIGUOUS REGULATION AND USE OF ULTRASOUND IN THE NETHERLANDS

If ambiguity is an intrinsic feature of ultrasound technology and its effects, its indistinctness as a technology certainly carries over into its (national) regulation and (local) implementation or use. An important question is whether ultrasound should be a standardized part of a prenatal screening program, or whether only suspicion of an abnormality should warrant the scanning of selected pregnancies. Giving pregnant women routine ultrasounds implies a collective decision to monitor a fetus's health, regardless of an expecting woman's medical condition. Women consider the prenatal exam to be a first acquaintance with their child-to-be and they are relatively unconcerned about its outcome.[31] Women who have a scan made because of a medical indication are usually worried about its outcome; the visualization of the fetus becomes a potential source of concern or grief.[32] Hence, we cannot separate national and local regulatory decisions from political and medical-ethical considerations preceding the official implementation.

Since the late 1970s, a number of prenatal tests have entered the market, each providing the pregnant woman with more information on the health status of her fetus. Besides ultrasound, women can opt for amniocentesis, serum alpha-fetoprotein screening, triple tests, and genetic testing for single-gene disorders. Every new test, of course, yields new information, and every additional bit of information may confront the pregnant woman with more options. Common sense holds that the availability of more tests allows expecting parents to make better choices. But prenatal tests that allow choices also require choices; norms often being as coercive as actual prescription, the very existence of a test affects parents' freedom to choose.

In many respects, the Dutch situation is peculiar. In most countries, first-, second-, and third-trimester ultrasound checkups have been routinely implemented as part of a standardized prenatal care trail; if additional scans are needed, a medical indication is required.[33] In the Netherlands, ultrasound is *not* a routine screening; theoretically, women *always* need a referral from their general practitioner or midwife to have a scan made in the hospital. The Dutch National Health Association explicitly considers routine screening in the first trimester to be unnecessary and inefficient.[34] Their regulatory policy is grounded on long-term research results indicating that routine screening has neither lowered fetal morbidity rates, nor has it proven to be cost-effective.[35] Medical scientists still disagree on the potential harmfulness of prolonged exposure to ultrasound; so far, experiments have indicated greater incidences of left-handedness, dyslexia, and lower birth weight as possible side effects.[36] In spite of a firm and clear national policy, however, first-trimester ultrasound scanning has practically become a routine procedure. Virtually every pregnant woman in the Netherlands obtains a medical referral from a general practitioner or midwife to have a first scan made in the hospital.[37] Medical referrals commonly state broad indications such as "to determine projected date of delivery" or "biometric measurement." Insurance companies accept these obviously general referrals, and thus implicitly acknowledge the psychological or emotional functions of a first-trimester scan. The incommensurability between regulatory policy and routine practice is obviously a sign of cultural acceptance of the first sonogram's "convoluted meaning."

Yet the discrepancy between regulation and practice is even more striking when we consider the traditionally low-tech nature of the Dutch prenatal care trajectory. Unlike the surrounding European countries, the Netherlands has a remarkably high number of deliveries outside a hospital; over 40 percent of all deliveries still take place at home, with the help of a midwife.[38] Prenatal checkups of normal pregnancies are performed by midwives or by general practitioners, and, in typical cases, the only time a woman is sent to the hospital is to have a first-trimester ultrasound scan made. Midwives are allowed to perform early ultrasound checkups, but insurance companies do not reimburse costs, so most midwives and health centers have never invested in machines. Hence, OB-GYN specialists in the clinics have monopolized the ultrasound scan, based upon the official policy that a scan always needs a medical indication. In a generally low-tech and de-medicalized trajectory of pregnancy and childbirth, the clinical ultrasound seems an anomaly.

In the mid-1980s and early 1990s, motivated by studies indicating positive psychological effects from ultrasound scanning, gynecologists began to promote the first-trimester sonogram as a unique social event and family ritual.[39] In the Netherlands, gynecologists tried to familiarize patients with ultrasound by inviting the pregnant woman's partner and children to the exam, giving a guided tour of the fetus, and allowing considerable time for parental bonding and psychological reassurance. On top of that, most hospitals provided prospective parents with pictures or even videotapes of their unborn child; initially, taping the ultrasound exam was a free service, but gradually clients became more demanding. The videotape became a cultural artifact—the first home video of their offspring. In satisfying these needs, hospitals whetted the appetite for more and better pictures. Gynecologists could exploit regulatory and insurance leniency to "scan for pleasure" as far as first-trimester sonograms were concerned, but scans of fetuses older than twelve weeks were restricted to high-risk pregnancies.[40] Over time, the ambiguity of ultrasound thus translated into a slightly confusing hospital regulation: on the one hand, hospitals promoted psychological and social meanings of ultrasound, while on the other hand, they strictly held on to the exclusive medical meanings in the visual-diagnostic checkup.

If, as is the case in the Netherlands, only medical urgencies warrant second-trimester scans, it should come as no surprise that commercial ultrasound clinics quickly found a niche in the booming pregnancy and baby business. In the early 1990s, when "baby's first video" had become an indispensable part of prenatal material culture, "ultrasound-for-fun clinics" began to appear in almost every Dutch city. Sonographers had never been regulated by state policy, so anyone owning an ultrasound machine could start a business. Perhaps due to a combination of the rapidly evolving, unforeseen success in commercial ultrasound and the profession's occupying a regulatory no-man's land between photography and medical specialty, sonographers faced few restrictions in the establishment of ultrasound-for-fun clinics. This unregulated moment resembled the situation at the turn of the twentieth century described in chapter 5, when photographers, catering to popular demand to provide loved ones with pictures of the intimate body, opened "X-ray studios" to take X rays of customers' hands and feet. Ultrasound-for-fun clinics in the Netherlands serve exactly such a function: they accommodate a pregnant woman's need for a picture or (more commonly) an extended twenty-minute video of the fetus for the family archive. In most cases, fun-sonographers are midwives or general practitioners with general obstetrical knowledge, who are trained in surface scanning. Some pregnant women visit these clinics at regular intervals to document the growth of their fetus; sonographers take advantage of the pictorial qualities of ultrasound scanners to take cute pictures of fetal movements such as yawning or thumb-sucking. Ultrasound-for-fun clinics charge anywhere between $75 and $150 for a picture or videotape. The gynecologists and midwives I interviewed mockingly refer to these clinics as the "Kodak departments of prenatal care."

AN ATTEMPT AT MEDICAL PURIFICATION

Fifteen years after the implementation of ultrasound as a regular diagnostic in prenatal health care, the first-trimester scan has inevitably become a mixture of medical checkup, psychological or emotional event, and photographic ritual. The ambiguity of the technology and its effects, manifest in ambiguous policies and regulations, has recently prompted a reevaluation of the sonogram's status and a recalibration of its use. In 1998, the Dutch professional organization of gynecologists called for locally or regionally coordinated efforts to cut down on nonmedical use of ultrasound. Although they are partly themselves to blame for the initial promotion of the sonogram as a social ritual, OB-GYN specialists gradually became convinced that the hospital was not the most suitable place for ritual family gatherings and photo ops. Since 1999, hospitals have narrowed the window of eligibility for first-trimester scans and discontinued the practice of taping the ultrasound for prospective parents. Moreover, gynecologists were willing to give up their long-savored monopoly on prenatal ultrasound, and agreed with insurers and policy-makers that beginning in 2000, midwives would be allowed to take over—and be reimbursed for—first-trimester ultrasound scans on healthy women with normal pregnancies. Not only did gynecologists consider the hospital too expensive a place for routine visual pregnancy tests and family rituals, their primary interest was to regauge the medical meaning of the ultrasound exam. The first-trimester scan, according to OB-GYN specialists, rarely shows any evidence of pathologies, whereas with a second-trimester scan, ultrasound technology allows for more accurate diagnosis and prognoses. In other words, the first-trimester scan, which is now adulterated with social, psychological, and cultural meanings, may be regulated outside the hospital, and the second-trimester scan has been promoted to the status of a "purely medical" ultrasound exam.[41]

The gradually changed meaning of the sonogram and the ultrasound exam appears to have resulted in a "natural" division of labor among Dutch health-care professionals, a division that could theoretically be translated into official policies. OB-GYN specialists in the hospitals perform ultrasounds for medical reasons only; their patients always carry medical referrals from general practitioners or midwives. Midwives perform first-trimester scans as part of a regular checkup, balancing medical monitoring with psychological and emotional needs. A prenatal health center is much cheaper than a hospital, and midwives already have extensive experience in counseling about early pregnancy loss and miscarriage, and can be trained in basic ultrasound scanning techniques. Finally, commercial ultrasound-for-fun clinics satisfy the client's desire for photographic evidence. The rationale behind this new division of labor is that, if every profession sticks to its business, the regulation of ultrasound will be medically justified, efficient, and will cater to the customer's need. The functions of doctor, psychologist, and photographer appear to be distinctly separate in this rational categorization.

However, growing practices are often unruly and professions may not be as easily classifiable as they appear. First of all, fun-sonographers do not see and promote themselves as simply photographers. When advertising their services, these clinics present themselves as semimedical centers that support the work of midwives and gynecologists; they claim to provide "extensive ultrasound prenatal exams" and check for "fetal morphologies and movements."[42] Advertising flyers usually boast the clinical qualifications of sonographers, and make it seem as if the exam provides an extra checkup, outside the regular medical circuit. Inevitably, fun-sonographers occasionally find a fetal defect—after all, they have commercially invested in rather expensive machines, and several years of scanning experience allows sonographers to build up visual expertise, even if they only have a general medical background. However sharp their observation, though, they are unable to adequately interpret a perceived visual aberration on a scan, and if they detect an irregularity, they cannot officially refer a pregnant woman to the hospital. The unofficial way to handle these situations is for a commercial sonographer to call a midwife in their personal network and have her write a referral for the woman to see a specialist. The gynecologists I interviewed generally find this practice annoying: more often than not, they are confronted with a false alarm—only an OB-GYN specialized in sonography can actually interpret the visual clues—but they are morally and emotionally obliged to perform another scan, if only to reassure the concerned client. Needless to say, there have been instances where a fun-sonographer has actually discovered a serious fetal defect, one that would have gone unnoticed if the woman had not decided to go in for a commercial scan.

Secondly, the continual advancement of technology stymies an easy reallocation of professional skills. Whereas ultrasound machines once could not detect certain fetal anomalies in the first trimester, the newest generation of scanners shows defects that might previously have been detected only later in the pregnancy.[43] Each innovation introduces technological improvements, upsetting the neat division of labor between specialists and midwives. Technology that is constantly in flux subverts the redistribution of power and expertise. Finally, the idea of reshuffling professional skills assumes a clear division between medical and social skills. Medical specialists are primarily trained to develop rational knowledge, interpretative skills, and technical dexterity. Yet as equipment and technical skills become more sophisticated, medical specialists have to become equally skilled at therapeutic and psychological counseling. An instrument that shows potential anomalies at an earlier stage of a pregnancy results in proportionally more anxieties, uncertain diagnoses, ambiguous pictures. The availability of ultrasound in combination with a host of other diagnostic tests—particularly genetic screening tests—requires a precarious mixture of high-tech knowledge, social skills, and empathy. Whereas medical professionals called for a rationalization of the medical deployment of ultrasound, the dynamics between technology, regulation, and humans thwarts that very same purification process.

Doctors who call for the "medical purification" of the ultrasound scan seem to want to cut the historical knot of denotations and connotations by reallocating meanings to distinct professional domains. In the above description of the ultrasound's implementation in the prenatal health care trajectory, I contest the view that medical and nonmedical meanings can be separated at all. In the course of fifteen years, the ultrasound pregnancy exam in the Netherlands has metamorphosed into a practice in which medical diagnostics, psychoemotional effects, and material-representational needs are inextricably intertwined. The expression "to have an ultrasound" concurrently signifies the diagnostic test, the artifact, and the ritual; hence, ultrasound is a complex medical-social-cultural phenomenon. Shifts in the regulation of professional skills and meanings often represent a scuffle to distinguish the boundaries between medicine and culture.[44] Indeed, ultrasound, like other medical imaging techniques, is an ambiguous instrument deployed in a hybrid practice; the convoluted attempt to purify that practice reminds us of the boisterousness of culture. Or, as Janelle Taylor aptly sums up: "The ultrasound exam highlights the fact that medicine is not a completely distinct domain of knowledge and practice . . . [but] involves struggles to establish clear boundaries between medicine and the broader consumer culture within which it is located."[45] The ambiguity of ultrasound technology and its hybrid effects are reflected in, and constructed by, local and national struggles for the professional authority to establish a hierarchy of meanings. The case study of ultrasound in the Netherlands provides a concrete illustration of such a boundary contest.

That contest is far from over, and neither is it restricted to the Netherlands. With the recent introduction of 3-D ultrasound technology, we can see the beginnings of a new regulatory struggle for professional authority. Commercial "fetal photo studios" are emerging in shopping malls around the United States, offering three-dimensional ultrasound sessions that are recorded on videocassettes or DVDs, complete with background music.[46] They are commonly performed after the second trimester, to obtain the most photogenic, realistic images of "babies yawning in the womb"; the package may also include picture-postcards, to be sent around as announcements to friends and family. Although the 3-D ultrasounds are advertised as "nonmedical," professional medical organizations as well as the Food and Drug Administration (FDA) have serious objections to this new practice. They warn that ultrasounds performed outside certified clinics may pose health risks and that unnecessary exposure to high-frequency sound waves may be unhealthy for the fetus. The competition between medical and cultural meanings of ultrasound has entered a new chapter, yet the same ambiguities underpin the nature of this debate.

MEDICAL INFORMATION AND THE AUTONOMOUS SUBJECT

As I stated earlier, there is more at stake in this local discussion than a mere regulatory struggle. The very assumption that sonograms can be cleansed of their social, cultural, and emotional "contaminations" implies that ultrasound technology principally yields objective medical information on the health status of the fetus. If we accept this logic, we contend that every prenatal test produces neutral clinical information, enabling a woman and her partner to make informed decisions, that the availability of medically objective facts allegedly precedes subjective, emotional choices.[47] Yet if I take seriously my previous conjecture that medical meanings can never be completely separated from nonmedical connotations, it is impossible to accept the premise of neutral medical information. Medical technology cannot be uncoupled from the way it is deployed, implemented, and regulated. The variety of scanning techniques, the different qualifications of sonographers, and the quality of the equipment all shape the scan's outcome; the very status of ultrasound—either as a routine screening or as a medically indicated diagnostic—already determines part of its impact and meaning. Moreover, the interpretive skills of gynecologists, midwives, and ultrasound photographers, even if they partly overlap, are also distinctly different. In sum, the premise that technology merely enables choices, that it provides neutral information preceding a pregnant woman's informed decision, seems hardly tenable. Choices are already partly anchored in the technology itself and in the way it is implemented in medical and social practice.

The idea that doctors provide medical information upon which a pregnant woman can base her decision also betrays a belief in the autonomous rational subject. Any pregnant woman submitting herself to an ultrasound exam, though, becomes part of a sociomedical trajectory. Medical experts and technicians play a normative role in this trajectory because they have, to some extent, standardized and institutionalized the meanings of sonograms. Beyond the hospital, family and friends impose norms and values—psychological or social pressure to have a fetus photographed reinforces the pictorial meaning and emotional value of the sonogram. The ultrasound scan's connotation as baby's first picture can hardly be forgotten when the woman comes back for a second scan and the gynecologist discovers a serious defect. However "pure" doctors want the ultrasound to be, medical information is inextricably bound up with experiential perceptions and relational knowledge from the very start.[48] Information, machine, pregnant body, and interpreting eyes are all interdependent. On top of that, medical scans often yield uncertainties rather than clear-cut rational calculations; complex probability statements provided by machines and doctors are considered rational knowledge, even if the information is unwieldy and difficult to manage. Complicated mediated knowledge forces patients to assess the information from an equally rational plane; the interpretations provided by doctors weigh heavily in the decision process, if only because many patients lack the technological expertise to understand the meanings inscribed in prenatal tests. Autonomy is a compromised notion in a medical context that increasingly requires patients to be highly educated problem-solving strategists.

Implicitly, the assumption of neutral clinical information and autonomous subjects is grounded in a gendered social hierarchy: whereas rationality is attributed to medical specialists, commonly men, experiential knowledge and subjective choice is ultimately a woman's business. The availability of prenatal diagnostic tests is often equated to the availability of legalized abortion and a woman's right to choose. I am not in any sense opposing better and more readily available prenatal tests. My point is, rather, that more tests do not in and of themselves advance the autonomy of female patients. In many ways, the belief in the pregnant woman as "autonomous" subject is as fallacious as the belief in the pregnant woman as a "victim" or object of the medical establishment and of technology. Writing in a different context about women's position as subject in reproductive medicine, Charis Cussins suggests that the use of medical technology does not automatically entail the objectification of the patient, and neither does it provide a clinical, neutral service to an independent subject; instead, recognition of the interdependence between technology, subject, and clinician urges pregnant women to think of themselves as active agents who "do not reject science and technology but try to negotiate a critical politics in use and development."[49] As my example of the specific situation in the Netherlands suggests, looking at sonograms increasingly involves an awareness of who is looking, who is being looked at, and in what circumstances.

Ultrasound's meanings are rarely exclusively medical. To accept ultrasound as a neutral information tool—as a purely medical diagnostic—precludes the notion of an embodied subject, who is neither objectified by medical practices nor unambiguously empowered by it, but who actively partakes in an intricate technological process that is also profoundly social and cultural.[50] Pregnant women have to be constantly aware of the fact that their choices are partly determined by the technological culture that surrounds them, and are partly the result of their own awareness of (and involvement in) clinical-technological processes. Prenatal ultrasound scans, at least in the Dutch example, are firmly embedded in a culture that initially promoted their psychological and social import; now that there are economical or professional reasons for the purification of their medical meaning, it makes sense to spell out the assumptions informing the promises of rationality and transparency. Decision-making processes concerning the continuation or termination of pregnancy are not rendered more transparent or rational when physicians have more and better imaging technologies at their disposal, and neither do they allow the patient more "freedom to choose." In some respects, the availability of more and more advanced prenatal tests—ultrasound figuring prominently among them—renders choices more complex and pregnant women less autonomous. High-tech probability statements make pregnant women more dependent on the judgments, knowledge, and technical expertise of specialists. The proposed separation of medical and nonmedical meanings enhances the authority of medical-technological rationality, even if this knowledge is rife with ambiguity and already inscribed with other than medical meanings.

The assumption that pregnant women and their partners can first assess medical information and subsequently make an informed decision is thus a theoretical bifurcation that can only be sustained in television drama. Or perhaps not? The *ER* episode following Peter Benton's first confrontation with the fetus on the ultrasound monitor dramatizes the premature birth of his son Reese, whose stay at the neonatal intensive care unit is only the beginning of a long trial of worry and anxiety for the parents. The neonatal lung-and-heart expert keeps Peter and Carla minutely informed of their infant's health status; born twenty-eight weeks into gestation, tiny Reese has to fight for his life. But the clinical information provided by the specialist, rather than helping or consoling the nervous parents, is so complicated and layered that it drives them to despair. Artificial respiration may instantaneously enhance the baby's weak breathing function, but in the long run, the use of the machine generates a 70 percent chance of permanent lung damage. Certain symptoms indicate potential brain damage or perhaps deafness; more testing may give more certainty, but may also cause new complications. Even Peter, himself a surgical resident, is clearly befuddled by the infinite stream of potentialities, uncertainties, and percentile predictions. When the medical expert has poured out his list of medical options for the novice parents, Carla desperately looks at Peter as if to say, "You are a doctor, so please decide what we should do." The usually rational and levelheaded Peter exclaims, utterly confused, "But that does not mean I know what is best for my son!"

CHAPTER 7

DIGITAL CADAVERS AND VIRTUAL DISSECTION

ANATOMICAL DISSECTION IS CONSIDERED AN ESSENTIAL INGREDIENT OF MEDICAL TRAINING. BY LOOKING AT AND CUTTING INTO DEAD BODIES, FUTURE DOCTORS LEARN TO DISTINGUISH BETWEEN HEALTHY AND DISEASED TISSUE IN LIVING BODIES, AND ALSO GAIN AN UNDERSTANDING OF THE THREE-DIMENSIONAL SHAPES OF ORGANS, VEINS, AND BONES. "ANATOMICAL DISSECTION" LITERALLY MEANS "SEPARATING THE BODY INTO PIECES"; THIS SYSTEMATIC DISASSEMBLING OF THE PHYSICAL BODY IS JUSTIFIED BECAUSE IT RESULTS in an entirely new body—a body of knowledge.[1] The confrontation with human cadavers functions as an important initiation rite for medical students: not until they have familiarized themselves with the face of death can they embark on the long educational journey that ends with a solemn dedication to life—the Hippocratic oath. Anatomy, from the outset, has been surrounded by sacral and secular symbolism; to this very day, the medical specialty has a morbid public image, associated as it is with the smell of decay and the aura of death.

Cadaver dissection does not provide the only occasion for medical students to become acquainted with organic human architecture. Anatomical illustrations help them conceptualize the forms and structures of various organs before they actually touch them. Without these two-dimensional representations, a thorough understanding of the body's physiology would be inconceivable. Ever since the sixteenth century, knowledge derived from close observation of cut-up cadavers has been recorded in drawings and anatomical atlases.[2] To convey their empirical findings, anatomists depended on the precision and craft of their illustrators. Accordingly, anatomical illustration is commonly viewed as mediated knowledge. Even the most sophisticated anatomical drawings, like those by Leonardo da Vinci and Andreas Vesalius, were considered derivative—idealized representations of real bodies.

From the early days of anatomy, then, anatomical training has been based on dissecting bodies and studying anatomical illustrations. In the 1990s scientists developed a new instruction tool that will purportedly revolutionize anatomy.[3] Funded by the U.S. Congress and the National Science Foundation, the Center for Human Simulation (CHS) at the University of Colorado, Boulder, created the first complete digital database of a human cadaver: The Visible Human Project (VHP).[4]

Supporters hail the VHP data sets as the ultimate replacement for both anatomical dissection and illustration in medical training, and as perfect simulations of real cadavers, neither disfigured by the anatomist's knife nor distorted by the artist's pencil. Digitization, the VHP implicitly states, offers eternal cadavers that can be "logged onto" regardless of time and place; two digital bodies constitute the necessary material to train future generations of students in anatomical practice and computer-assisted surgery.

The production of a complete virtual body involves a range of complicated, state-of-the-art techniques. First, the body needs to be digitized by means of magnetic resonance imaging and computer tomography. Then the cadaver is immersed in a special fluid and deep-frozen to minus seventy degrees Celsius. Next, lab workers use a precision planing device, a cryogenic macrotome, to shave off millimeter-thin slices of the body. After each slice, the cross-sected profile is photographed digitally and scanned into a computer. The resulting data set constitutes the basis for an unlimited series of three-dimensional simulations. In 1995, the digital body of the Visible Male became available for public use, followed one year later by the even more detailed data set of the Visible Female. Both data sets are accessible through the National Library of Medicine's Internet site.[5]

The VHP has presented the digitized Visible Human as a revolution in anatomy: Since computers enable three-dimensional images of bodies, the old two-dimensional representations have become outmoded; moreover, three-dimensional images may render dissections of real bodies obsolete. According to VHP director Michael Ackerman, the creation of digital cadavers signals a radical break with traditional anatomical education.[6] However, I argue that virtual dissection and digital cadavers constitute a distinct continuation of age-old anatomical practices. The VHP is firmly rooted in historical conceptions of the body and its representation, crime and punishment, and physiology and art. Rather than breaking with educational traditions, the VHP reflects a renaissance of public anatomy lessons. In order to develop this argument, we have to return to Renaissance Europe, where the first public anatomy lessons took place, and where, slightly later on, in cities like Padova, Bologna, Amsterdam, and London, special anatomical theaters were built to accommodate the large crowds these lessons attracted. In those days, public dissections not only served educational purposes but were linked with the criminal justice system and used to teach moral lessons. In this way, the early anatomical theater served three institutional functions: school, court of law, and public spectacle. A close consideration of these three functions will unearth the VHP's cultural roots; it will reveal how this American digital project, that in many ways seems so characteristic of our postmodern day and age, is a directly descendant of European Renaissance tradition.

THE ANATOMICAL THEATER AS A SCHOOL

The first and foremost goal of the Visible Human Project is educational: it should help medical students to become better doctors. Digital anatomy, VHP director Michael Ackerman posits, has considerable advantages over both conventional dissection and anatomical illustration.[7] The Visible Human, he claims, offers a standard model for twenty-first-century anatomy, enabling students to obtain valuable clinical training without the need for prepared cadavers. Real cadavers are expensive, perishable, and commonly tainted by disease. Moreover, cutting into dead bodies is an uncomfortable experience for many students. Anatomical illustrations for their part reduce anatomical structures to two-dimensional flat surfaces and are therefore insufficient teaching tools in and of themselves. The Visible Human purportedly improves on two-dimensional representations because digitization allows viewers a three-dimensional perspective on body parts and organs. Before qualifying these claims, I need to elaborate on the role of anatomical dissection in the history of medical education.

Cadaver dissection as part of the medical curriculum dates back to the fifteenth century, but historian Katherine Park firmly rejects the myth that cutting into dead bodies rarely happened before that time.[8] For instance, in Italy dissections are recorded as early as 1286. Autopsies to detect unknown causes of death formed a regular part of medical practice, and postmortems are known to have taken place in the early fourteenth century at the University of Bologna's medical school. Dissection was sometimes warranted by suppositions of sainthood: a recently deceased person's body would be cut open in the hope of finding physical signs and symbols of sacredness.[9] The educational value of postmortems and autopsies, as Park contends, was still rather limited. University professors taught their students the principles of human anatomy by dissecting cadavers, yet dissection was merely an extension of anatomical illustration: "Their goal was not to add to the existing body of knowledge concerning human anatomy and physiology but to help students and doctors understand and remember the texts in which that knowledge was enclosed."[10] As we can tell from fourteenth-century depictions of anatomical lessons, the transmission of knowledge was hierarchically structured.[11] Galen's anatomical and physiological theories had been accepted since the fourth century as the highest authority in medicine. His texts were read ex cathedra by a professor or lector while an *ostensor* pointed to the organs in the body laid open on the table. The actual dissection was left to relatively unimportant menials, dissectors who were considerably lower in status than the lector or ostensor because the demonstration of dissected body parts played a minor, purely ornamental role.

The sixteenth century was the height of public anatomical dissection. The Flemish anatomist Andreas Vesalius upset the assumed hierarchy between anatomical theory and manual anatomical skill.[12] He noted errors in Galen's anatomical descriptions and relied, instead, on empirical evidence.[13] Vesalius's view and practice were not only an unscrupulous sacrilege of Galen but an outright condemnation of Galen's followers and their uncritical acceptance of his theories—that is, without empirically verifying them. Vesalius simultaneously performed the acts of dissection and explication, undermining the authority of the text by prioritizing the

Fig. 18. Anatomical Theatre, Leiden, Netherlands. Boerhaave Museum.

tactile dimensions of the body.[14] By the time the British anatomist William Harvey was conducting his public anatomies, it was common practice for anatomists to handle both instruction and dissection.[15] Hands-on contact with the cadaver became the exclusive privilege of the anatomist. During public dissections, students were not allowed to touch any body parts; only afterwards, behind closed doors and in small sessions, were they given the opportunity to test their manual skills. The anatomist's personality and oratorical talent largely determined the educational value of public lessons, but it is safe to say that they were of little instructional use to anyone except, perhaps, the anatomist himself.

The anatomist, rather than the cadaver, constituted the focal point of the anatomical theater. It was his task to lead the public from the observation of a single dead body to abstract theories about living bodies. Some anatomists proved to be excellent performers, and they were capable of convincingly translating their concrete tactile and visual perceptions into imaginative oratory. But without expert explanation there was little for students and the general public to learn. The sixteenth-century anatomist functioned as a synecdoche for the medical corps, or "body of knowledge." Although the dissecting professor literally visualized physiology by exhibiting body parts and viscera to the crowds, only those who sat up front, close to the dissection table, could actually observe his operations. Those further in the back had to rely on the anatomist's verbal explications, which were comprehensible only to the initiated. For the average spectator, who was generally illiterate, the anatomist's elucidation in Latin did not add anything to the visual demonstration.[16]

The Visible Human Project is conceived of as a simulation of anatomical dissection—or, more precisely, as its emulation in a virtual environment. Digital cadavers, which can be downloaded from the Web, will return anatomy to the public eye and reintegrate (and reshuffle) the dimensions of learning, seeing, and touching. According to Michael Ackerman, a prime reason for creating digital cadavers is that "much of our understanding of complicated health and disease processes actually lies in images, not in text."[17] Three-dimensional reconstructions of real cadavers enable medical students to connect theoretical and empirical knowledge. By recompiling Visible Human data into gross anatomical structures, students may look at a large number of cross sections and spatial representations of organs or body parts. As a surgeon examines radiological information on the screen, students could perform virtual dissections to get a better sense of anatomical structure. At a later stage, the Center for Human Simulation will create a virtual surgical unit, complete with radiological and anesthesiological simulators. The virtual body, including blood circulation and physiological reflexes, will help train future surgeons in the quintessence of medical practice: live surgery.[18] The simulation units will resemble virtual cockpits designed for Starfighter pilots, offering hands-on experience without endangering real people's lives. The term "hands-on" suggests that virtual dissection is seen as a perfect replacement for the tactile experience obtained by regular dissection; manipulating digital pictures with a computer mouse is seen as the equivalent of handling a dissector's knife.

The cultural history of anatomical dissection, however, invites us to put the overtly ambitious claims advanced by the VHP into proper perspective. In the early sixteenth century, Andreas Vesalius challenged the dominant Galenic paradigm by putting practice before text, but at the end of the twentieth century, tactile experience is on the verge of being replaced by the visual. Text, image, and body may coalesce in a digital cadaver, yet in the hierarchy of senses, image clearly dominates. The VHP presents itself as a visual reference book—a standard body for the twenty-first century. However, it is rather presumptuous to equate digital reconstructions with actual cadavers, and to consider virtual dissection as valid as its real-life counterpart. Nevertheless, this does not mean that working with the VHP's data sets constitutes a lesser or less valuable preparatory tool for future doctors. On the contrary, getting acquainted with human anatomy through digital cross sections perfectly suits contemporary medical practice, which already relies heavily on scanned cross sections of patients' bodies. The particular significance of the Visible Human for medical education, then, may be less in the digitization of human cadavers than in the digitization of medicine as a whole. Increasingly, the contemporary hospital is a digital environment. Since specialties like surgery or oncology have become almost fully dependent on digital representation technologies for diagnosis and treatment, medical training relies more and more on looking at human bodies by means of computers.[19] CT and MRI scans determine the specialist's view on the body, and reading cross sections of the anatomical body is completely commensurable with the practice of reading the living body. What virtual anatomy has done to the field is to bring anatomical practice back to the frontiers of modern, digitized medicine, and thus to the spotlight of public attention.

The superiority of anatomical instruction via three-dimensional representations to actual dissection is hard to substantiate. In the Renaissance, the public anatomy lesson was mostly instructive for the anatomist himself, and was of questionable use to students not already versed in the prevailing anatomical theories, let alone to a general public. By the same token, digital dissection has true educational value only for those who already have a basic knowledge of anatomy. Digital cross sections of the Visible Male mean as much to a layperson as the anatomist's Latin explications meant to the sixteenth-century spectator. For one thing, the data set itself is nothing but a series of bits and bytes, and it takes competent medical software specialists as well as considerable computer storage space in order to translate and recompile those data into usable fly-throughs. Moreover, interpretating cross sections requires substantial experience. The big difference between virtual and conventional dissection is that with virtual dissection the body is conceptualized in slices rather than pieces; the less-than-a-millimeter slices mean little or nothing if the viewer cannot relate these slices to actual three-dimensional organs. Just as decoding ultrasound images or X rays requires training, one needs an experienced eye to translate MRI and CT scans into anatomical structures. To learn anything at all from a virtual dissection, a student already needs a basic competence in anatomy. Since virtual anatomy instruction is visual rather than tactile, "eyes-on experience" seems a better term for virtual anatomy than the "hands-on experience" it advertises. Needless to say, the smell of decay and the aura of death are completely absent from the virtual three-dimensional learning environment.

Besides the bold promise that the Visible Human will replace conventional dissection in medical training, its proponents claim that digitized cadavers emulate classical anatomical illustrations. In the Renaissance, artists often closely observed dissections and subsequently translated the anatomist's knowledge into detailed drawings of the body's interior. Because anatomists were rarely gifted illustrators, lectors depended on the precision of their illustrators for a faithful and accurate depiction of the anatomy. Atlases commonly resulted from this close collaboration between artist and anatomist.[20] Like floras and geographical atlases, anatomical atlases and illustrations have always been the result of a collaborative effort between art and science. Yet artistic interference detracted attention from physiology to art, from the scientist to the artist. In the eyes of many (the initiators of the VHP among them) this mode of representation has two major drawbacks: it has led to idealized, abstracted, and often "distorted" representations of the human body, and, most importantly, it always has involved projection of three-dimensional structures on a flat surface.

By contrast, the digital data sets offer three-dimensional images that can be rotated so that projected body parts may be seen from any plane or perspective. On the computer screen, students can manipulate the images with a mouse, and pre-modeled fly-throughs enable seamless crossovers between organs.[21] A digital cadaver, the VHP claims, is no longer a *representation,* tainted by the subjective interpretation of an artist, but a *simulation*—a digital reconstruction of a real body. The VHP data sets, which are increasingly used as a basis for drawings, have allegedly brought anatomical illustrators away from art and closer to science.[22] Students no longer have to depend on anatomists to translate tactile experience into words or artists to translate observation into illustration. The VHP suggests that its digital simulations constitute "unmediated inscriptions" of cadavers that are distorted neither by the pencil of the illustrator nor by the knife of the dissector.

Asserting that digital inscription is beyond representation, however, seems exaggerated, if not unwarranted.[23] The introduction of a new visualizing technique, starting with the X ray in 1895, has in fact always been accompanied by the enthusiastic claim of increased transparency. But in each case, the claim has proven to be illusory. The application of every single new technique—whether CT, MRI, or digital photography—results in a specific new focus on the body. Even a combination of all available perspectives will never produce an undistorted, transparent body. Invariably, our view of the body is informed by the modality of its visualizing instruments. We cannot fully comprehend the virtual body without a preliminary understanding of both anatomy and its representational techniques. In the educational setting of anatomy instruction, knowledge of visualizing technologies is as important as knowledge of anatomy itself. The Visible Human is not beyond representation because digital imagery imitates body shapes better than conventional anatomical illustration; rather, the illusion of verisimilitude is primarily due to the fact that digital images are now a common visual currency in medical practice. Even though digital scans have quickly become indispensable tools in the art of anatomical illustration, they do not, in effect, eradicate artistic subjectivity.

Medical schools around the world have already incorporated the Visible Human into their curriculum, while discussions on whether digital dissection will replace or simply complement traditional cadaver dissection are ongoing.[24] Indeed, the digital environment of any modern-day hospital requires doctors to have rigorous training in imaging and interpretation. Yet without knowledge of, and hands-on experience with, the anatomical body—without text and touch—the three-dimensional images of the Visible Human lack the surplus value attributed to them.

THE ANATOMICAL THEATER AS A CRIMINAL COURT

Most people who donate their body to science after death regard their gift as a noble contribution, one that enables medical students to practice dissection, which in turn will help them save lives later on. The provenance of cadavers in Western teaching hospitals is regulated by strict protocols and is rarely a subject of discussion; as a rule, the cadaver's anonymity is guaranteed, so students can concentrate exclusively on the scientific dimension of the dead body. By the same token, scientific articles referring to the Visible Human omit any information regarding the creation of these databases. However, the provenance of these digital bodies—the actual bodies on which they are based—forms a crucial subtext for understanding the historical and cultural roots of the Visible Human Project.

Until the late fifteenth century, anatomists generally used unclaimed cadavers for dissection—the bodies of individuals who had died of some kind of disease and whose bodies were donated or left unclaimed,[25] usually because there were no relatives or friends to take care of their burial. The supply of unclaimed cadavers kept pace with the number of bodies needed for dissection each year, until around 1500, when public dissections began to attract larger crowds and the demand for fresh cadavers increased accordingly.[26] To keep up with demand, anatomists started to seek out the cadavers of convicted and executed criminals. For judicial, moral, and scientific reasons, anatomists preferred the bodies of executed criminals. In the early sixteenth century, public dissection became directly connected to the criminal justice system when the courts wielded it as a form of extra punishment on top of the death penalty. By the seventeenth century, public dissection was common practice in most European countries;[27] in Britain, it was explicitly incorporated in the famous Murder Act of 1752, which was designed to teach shameless bandits a moral lesson. Death by hanging or execution was considered too mild a deterrent to crime, but public dissection appeared to be a daunting punishment. After the humiliation of a public execution, the criminal's soul suffered a second indignity by means of a public dissection. After the criminal's execution, it was believed, his or her soul would be floating around the body for several more days. The executed criminal was denied a decent burial—a final resting place for the soul. The combined public execution and dissection constituted, as Jonathan Sawday aptly put it, "two acts in a single drama."[28]

From a disciplinary point of view, the anatomist functioned as an extension—sometimes even literally—of the executioner: both were in charge of executing the sentence imposed by the judge.[29] Although anatomists explicitly distanced themselves from ordinary executioners, they had a similar professional interest in capital punishment. It may not be a coincidence that the establishment of anatomical theaters often led to an increase in the number of death-row convictions.[30] Public dissection, like execution, worked as a moral deterrent; it gave spectators the sense that those who had harmed society were forced to give something in return—murderers in particular were sentenced to death plus dissection. Punishment in the service of medical science had considerable moral and symbolic value. Moreover, by paying his dues to society, the convicted criminal's chances of ending up in purgatory or even heaven increased considerably.[31]

For scientific uses, Renaissance anatomists preferred the corpses of executed criminals. This created an interesting paradox: cadavers had to be identified as criminals in order to function as a moral deterrent, yet anonymous in order to serve as a scientific object. In most of Europe it was common practice to safeguard the anonymity of a corpse for dissection; the British Murder Act even legally stipulated it. In contrast to public executions, which usually took place in the convicted criminal's hometown, public dissections were performed with cadavers brought in from nearby towns.[32] Out of respect for the criminal's family—to spare them the added shame—the cadaver's identity was not disclosed. But that body had to be identified in order to serve as a moral deterrent. The anatomist solved this dilemma by listing, at the beginning of the dissection, the crimes for which this body had been sentenced to death, so that the moral lesson was made explicit without the identity of the criminal being exposed. There was yet another reason for withholding the criminal's name: to emphasize the body's representativeness rather than its uniqueness. Identification of the body would have distracted from the scientific nature of the anatomy lesson, since the audience would see a dead criminal instead of a scientific object on the dissection table. After all, dissection was not meant to expose the interior of a particular corpse, but to extract general knowledge about the human body.

Although morally tainted, the cadaver had to be physically untainted. In contrast to "found" or donated bodies, which were commonly disease-ridden, the bodies of executed criminals were usually in good shape. It is known that anatomists told judges that they were specifically interested in bodies of average length, age, and size.[33] In order to be representative, the bodies used at dissections had to fit the audience's sense of what constituted a normal healthy body. The criminal's moral repugnancy was at least as important as his or her mint physical condition; but, although his crimes had to be utterly heinous in order to set a moral standard, his body had to be untainted by either disease or the execution of the death sentence.[34] As a rule, the judge granted the anatomist's request to have the criminal hanged rather than executed in another manner, because the gallows left fewer physical marks and disfigurements.

Female bodies were more scarce, as there were fewer women convicted. In general, female convicts were sentenced to the dissection table for smaller offenses than their male counterparts. Male criminals usually received capital punishment for murder; female criminals, for infanticide or theft. Female corpses offered an opportunity to demonstrate the female reproductive system, and the younger their bodies, the better. In a society where the veiled female body was associated with chastity and honor, and the naked female body with seduction and shame, female cadavers obviously provided an attraction for a mostly male audience.[35] In the gendered social order of the Renaissance, it should come as no surprise that female cadavers were held up against different standards of punishment and moral judgement from male cadavers.[36]

The cultural and historical ties between the medical and the judicial system, as discussed above, elucidate our understanding of contemporary virtual anatomy.

The VHP confronts us with similar paradoxes concerning the body's anonymity and representativeness. The identity of the Visible Male did not remain a secret for very long. When the CHS put the Visible Male's digital data on the Internet, they refused to disclose the identity of the man it was modeled after; the only fact they revealed was that he had been a thirty-nine-year-old Texas prisoner who had been sentenced to death. Yet, since the CHS had released the date of the prisoner's conviction, it was fair game for journalists to try to trace his identity. The "real" Visible Male turned out to be Joseph Paul Jernigan, who had been sentenced to death for murder and robbery on August 26, 1993. While still on death row, he had agreed to donate his body to science—specifically to the Visible Human Project. In exchange for his collaboration, his sentence of death by electric chair was changed to death by lethal injection. Just as Renaissance judges sentenced criminals to the gallows to please anatomists, the milder sentence issued to the Texan prisoner was primarily motivated by a desire to serve the interests of science: lethal injection causes minimal effects on the otherwise perfect body. In line with sixteenth-century European conventions, the body was not dissected in his home state, Texas, but transported to Colorado, where the cryogenic macrotome saw cut it into 1,872 thin slices.

Scientifically speaking, the identity of the man whose material body constituted the basis of the digital data set was absolutely irrelevant, if not harmful; therefore, the directors of the VHP were reluctant to release the Visible Male's identity. However, the ease with which journalists tracked down Jernigan's personal record calls this intention into question. The media instantly turned Jernigan into a posthumous celebrity; without the specific personal data, which was easily tracked down by journalists, the project would have most likely received half the media attention.[37] Looking closely at the VHP's media coverage, it is remarkable to see how Renaissance morality resonated in the 1995 newspaper clippings.[38] Jernigan's dissection into slices was described as "extra punishment" on top of his death sentence. Some newspapers commented that, by donating his body to this educational-scientific project, Jernigan had at least paid his dues to society. Other commentators appeared outraged that this murderer was granted eternal life on the Internet—virtual reanimation as a reward for his hideous crimes.[39] In whatever capacity, the mythology of the convicted criminal has become an integral part of the Visible Male data set; Jernigan's digital representation remains a constant reminder of a criminal body disciplined through capital punishment—first death by injection and then dissection—forever exposed to the scopic regime of science.

Jernigan's alleged representativeness of the average living body forms a second paradox, for the virtual cadaver undeniably exhibits some very distinct idiosyncratic features. According to the directors, finding a suitable body for the Visible Male proved rather difficult; it took them more than two years to find a body that could serve as a "standard for anatomy"—that is, a body of average size, length, weight, and height, with no visible physical imperfections. To be representative of the average living male, the model should not be too old, too tall, or too heavy, and should, ideally, lack any physical abnormalities. Joseph Jernigan was chosen from five other potential candidates because of his mint condition: a healthy 170-pound male who had lifted weights in prison on a daily basis. Yet his body, as it turns out, was not exactly perfect: besides a missing appendix, one testicle had been removed to prevent a benign tumor from growing. Apparently, these abnormalities did not disqualify his body as unrepresentative. On the other hand, some external features, such as a tattoo on his arm, were seemingly left clearly visible on the screen, to certify the body's "authenticity." The violence of execution and dissection—the lethal injection and the cryogenic dissection—left no visible signs on the digitized body. The virtual cadaver represents any living body, even though the material referent, Joseph Jernigan, seems anything but representative.

We may perceive similar paradoxes concerning anonymity and representativeness in the data set of the Visible Female. The Visible Male began his life as a shade a year before his female counterpart joined him on the Internet.[40] The Visible Female was not recruited from the circles of criminals but modeled after a fifty-nine-year old Maryland housewife who had signed a codicil to donate her body to science after death. Unlike Joseph Jernigan, she had not specifically intended her body to be used for the VHP, but her husband decided after her death that this was a noble and important cause. Her husband confirmed her apparent excellent condition by stating that his wife had never been sick a single day until she was struck by a heart attack. The female cadaver underwent the same treatment as Jernigan's, the only difference being that she was cut into slices of one-third of a millimeter, resulting in an even more refined database.[41]

In contrast with Jernigan's case, we know nothing about the woman's identity but her age and status; in the media attention that followed, the Visible Female was primarily evaluated on the basis of these gender-specific features.[42] The label "housewife" suggested the normalcy of the woman—one whose body is unaffected by intellectual or otherwise "untypical" female activities. Yet the woman's representativeness is clearly undermined by her age: since she is postmenopausal, her body cannot serve to illustrate the female reproductive functions, and the directors of the VHP have subsequently agreed they should make up for this deficiency by searching for a younger sample. Apparently, there are different standards for determining the normalcy of virtual males and females. Despite his missing testicle, Jernigan is still considered "standard," while the Visible Female is looked upon as substandard because she is no longer fertile. Even though menopause is not the pathological equivalent of a missing testicle, the Visible Female is not considered representative of the average female body. Significantly, the gender-specific cultural criteria that we use to differentiate between living men and women are unilaterally projected onto these virtual cadavers.

As was the Renaissance anatomical theater, the VHP is bound up with the criminal court system. Whereas formerly in Europe criminals were sentenced to death and dissection, today in the United States criminals are sentenced to death and asked if they mind being cross-sectioned. By confronting the audience of public anatomy lessons with the body's criminal past, moral content is added to the anatomical dissection. Similarly, in contemporary virtual anatomy, contextual information on the Visible Male's identity turned out to be a major factor in popularizing the project and disseminating its data sets. The major paradoxes that were prevalent in Renaissance anatomy resonate in the VHP. Neither the Visible Male nor the Visible Female truly reflects the ideal of transparency. Their creation is deeply embedded in concrete historical and cultural contexts, and in their organic materiality, their specific ties to individual bodies with their inescapable idiosyncrasies, they fail to be representative.

THE ANATOMICAL THEATER AS A PUBLIC SPECTACLE

In addition to its ties to the educational and justice systems, anatomical dissection was also an early form of mass entertainment—of public spectacle. Although this dimension of anatomical practice virtually disappeared after the late eighteenth century, it is surprising to find how the VHP has contributed to its return. In the sixteenth century, Vesalius's dissections, giving rise to an empirical turn in anatomy, induced a shift from private to public instruction. Most likely due to the change in emphasis from textual to personal authority, the audience of the anatomy lesson increased from a handful of students to large crowds of spectators attracted from all social strata. To accommodate large crowds, theaters built particularly for public anatomy lessons mushroomed in Europe during the seventeenth century, especially in towns and cities with universities that had medical faculties, such as Bologna, Padova, London, Leiden, and Amsterdam.[43] They became important

Fig. 19. Frontispiece of Andreas Vesalius's De Humani Corporis Fabrica, *1543.*

centers of innovation in the study of medicine. Famous anatomists like William Harvey, who was the first to explain the principle of blood circulation, attracted large groups of both medical students and laypersons. Architecturally, anatomical theaters were designed after typical Renaissance theaters, with round stages and gradually ascending seats enabling the audience to gaze into the cadaver from above.

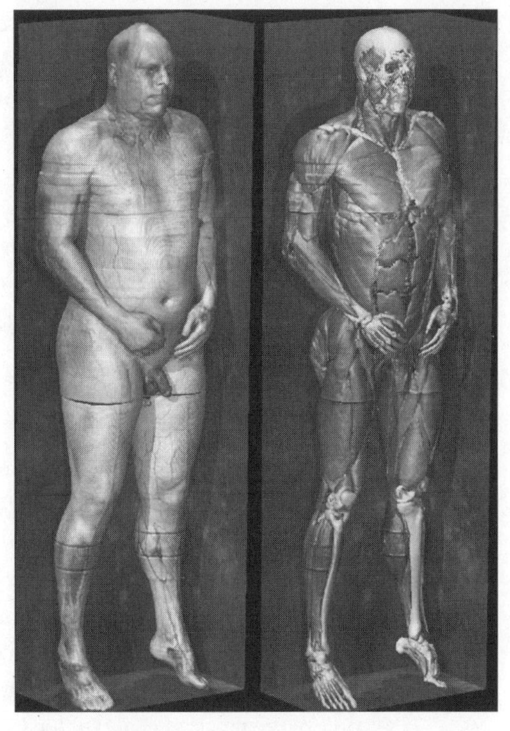

Fig. 20. VOXEL-MAN computer-based anatomy model. Torso with and without skin. Institute for Medical Informatics, University of Hamburg, Germany. Courtesy of Karl Heinz Höhne. www.uke.uni-hamburg.de/voxel-man.

Illustrations and drawings from Vesalius's famous anatomical atlas *De Humani Corporis Fabrica* show large and varied crowds surrounding the anatomist and the cadaver.[44] Skeletons and Vanitas symbols decorated the open space, underscoring the moralistic intentions of the public anatomy lesson. It was not uncommon that a banquet, concert, or other performance accompanied the dissection, resulting in an event that lasted up to several days. Dissections usually took place during the cold season, February being the most popular month. Obviously, the low temperatures helped conserve the cadaver for several days. Some historians have associated the spectacle of dissection with the annual carnival.[45] Tickets for anatomical festivals did not come cheap and were much in demand; we know from historical tracts that rituals and ceremonies added luster to anatomical lessons.[46]

The public anatomy spectacle was highly influenced by the conventions of the Renaissance morality play. In seventeenth-century dramas, such as those by Shakespeare in Britain and Vondel in Holland, catharsis or purification of the soul was a central element. Anatomy lessons, much like the Elizabethan tragedies that ended with the death of all of the characters, had a decidedly morbid tone, moving from a fresh and preferably perfectly intact corpse to what was basically no more than a gnawed-off skeleton. Bare skeletons on the walls of the anatomical theater foreshadowed the cadaver's inevitable fate. In his public lesson, the anatomist articulated the relation between the dead body on the table and the living body of the

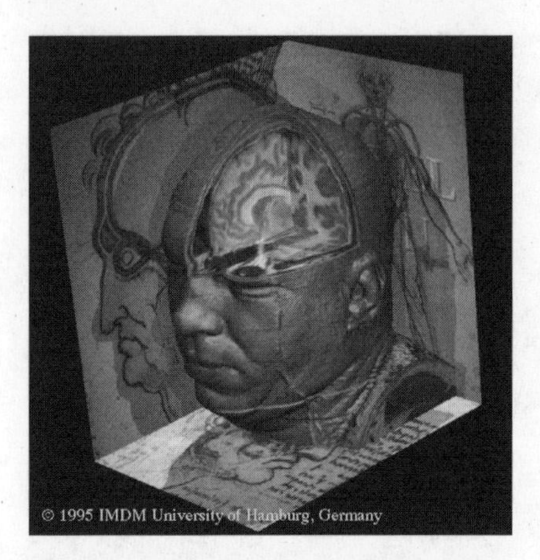

Fig. 21. VOXEL-MAN computer-based anatomy model. Head. Institute for Medical Informatics, University of Hamburg, Germany. Courtesy of Karl Heinz Höhne. www.uke.uni-hamburg.de/voxel-man.

spectator. Catharsis was only immanent if crime and punishment were theatrically related to the morality of health and medical science.

Anatomical theaters played a major role in acquainting a large audience with the advances of science. The popularity of anatomical theaters in the Renaissance should be regarded in the context of the rise of *Kunst-* and *Wunderkammer* and botanical gardens, which also exposed large crowds to the marvels of nature and science. However, in their public dissections, anatomists did not so much intend to share their knowledge with a general audience as to impress them and command respect and awe. As anatomical theaters flourished, they turned into cultural centers where scientists and artists worked side by side, inspiring one another. Many famous painters and writers (most notably Rembrandt) attended public dissections and recorded anatomical spectacles in their artwork. Scientists and artists used the same "raw material," only for different purposes. Although few artists really understood the Latin oracle in front of the cadaver, they were fascinated by the combination of scientific aura, moral lesson, and morbid entertainment.

The same cocktail of science, morality, and morbidity can also be retraced in the Visible Human Project. This expensive virtual anatomy project was evidently designed for educational purposes, but after its results were released on the Internet, the data became public property and have—perhaps inadvertently—already been used by a variety of professionals for very different purposes.[47] For instance, products derived from the Visible Male and Visible Female's data sets include sophisticated interactive fly-through animations, such as the one developed by the State University of New York at Stony Brook for the detection of bone-marrow cancer. The University of Hamburg initiated the VOXEL-MAN Project, turning the data into a number of specific digital fly-throughs of organs and body parts for purposes of education, medical illustration, and promotion.[48] And, just as in the Renaissance, artists have used the Visible Human data sets to create their own versions of corporeal imaging for artistic and entertainment purposes.

When we take a closer look at one of these popularized products, it is easy to find historical echoes of the anatomical theater's function as a public spectacle. *Body Voyage: A Three-Dimensional Tour of a Real Human Body,* a CD-ROM with a companion book, vaguely pretends to disseminate anatomical knowledge to a lay audience, but its educational value proves very minimal indeed.[49] Except for a few general medical facts and figures, this CD-ROM contains no serious information on either physiology or anatomy. The user has to understand and interpret colorful cross sections without any guidance. For laypersons with little prior knowledge of physiology, the educational value of this product is close to zero. The high entertainment value of *Body Voyage* is undoubtedly due to its spectacular presentation and dramatic plotting: the recompiled data set of the Visible Male is presented as the gloomy "inside experience" of a dead criminal's body. In addition to the minimal scientific content, the only explanations we get from scrolling through the disk are juicy details about Jernigan's background and criminal past. We learn how he was sentenced to death after robbing and killing a seventy-five-year-old man, and that he never showed any remorse for his cold-blooded crime. The CD-ROM minutely describes the details of Jernigan's execution, including his last words and information on his last meal (a cheeseburger). The dissection device, a cryogenic macrotome, is shown on a QuickTime video display. As did the anatomical theater, *Body Voyage* capitalizes on the spectacle of dissection: the spectator can virtually dissect Jernigan's cadaver layer by layer. The software channels the viewer's gaze into the cadaver, where he is able to execute the dissection "hands on" by actively "cutting up" the body with a few simple mouse clicks. The CD-ROM displays the drama of the criminal who deserved to be sentenced to death and dissection. The story clearly aims at bringing about the desired catharsis, entertaining yet disciplining the spectator, drawing him into the virtual experience of crime and punishment.

The artist who made this CD-ROM, Alexander Tsiaras, views it as an artistic interpretation of the Visible Male's data set. For the production of this CD-ROM, Tsiaras used some of the digital data provided by the CHS, yet retouched their original colors in order to create a more aesthetically pleasing digital cadaver. Tsiaras places himself in the tradition of great artists like Leonardo da Vinci, Albrecht Dürer, and Rembrandt, and his explicit purpose is to "inform and entertain a popular audience." As stated in its introduction, *Body Voyage* is the result of a perfect marriage between science and art. It is the combination of colorful cross sections, suspenseful plot, a "truly" horrendous character, and morality play that warrants the success of this product. Rather than serving as an instructional tool, this derivative of the VHP mainly echoes the Renaissance anatomy lesson's function as public spectacle. Disguised as popular science education, the CD-ROM invites the user to subject the convicted and executed prisoner, time and again, to virtual dissection.

DIGITAL CADAVERS AND VIRTUAL DISSECTION

Considering the Visible Human Project in light of the various functions of the anatomical theater, it represents a distinct continuation of historical anatomical practices. In virtual anatomy, just as in Renaissance theaters, the disciplining of the living body takes place through the dead body. However, despite the various continuities, the VHP also suggests that there are clear discontinuities with sixteenth-century anatomical practices. For instance, virtual anatomical dissection takes place in the context of individualized computer education rather than in public theaters, and the difference between these two is considerable. Moreover, Joseph Jernigan's actual dissection occurred within a different sociopenal code, and we should not ignore the fact that, within the constraints of the death penalty, being dissected by the cryogenic saw was Jernigan's own choice. It should be emphasized that I do not believe that the VHP directors made a conscious effort to situate digital dissection in a European Renaissance tradition. My goal was merely to elucidate how the current practice of compiling and disseminating digital body data reflects and constructs persistent cultural norms involving age, gender, spectacle, identity, transparency, and crime and punishment, and how these various norms are historically interrelated.

The arguments bolstering the claim that the VHP will replace both conventional dissection and representations in anatomical atlases seems overstated. Virtual dissection certainly offers excellent training opportunities for future doctors. It is also true that Jernigan and the Maryland housewife have become, in some respects, immortal and imperishable as shades on the Internet. But whether the Visible Male and Female serve as universally representative cadavers is highly questionable; despite the VHP's attempt to move beyond idiosyncrasy, both bodies are distinctly inscribed by cultural and social norms. Moreover, scrolling with a computer mouse is hardly the digital equivalent of actual dissection, let alone an emulation of it. Although surgeons increasingly rely on computers and digital technology in general, as long as the empirical body remains the focal point of clinical practice, a digital cross section can never replace anatomical dissection.[50]

Digital dissection may seem less inscribed with sacral symbolism and morbid connotations than conventional dissection, but the "new standard of human anatomy" hardly consists of a transparent set of digital data.[51] As we saw in chapter 3 in relation to Gunther von Hagens's Bodyworlds, the material basis of anatomical models and their representations, to a large extent, mirror contemporary standards of acceptability and aesthetic preferences for certain modes of display. Like the plastinated corpses, the Visible Male and Female tell us as much about our contemporary cultural tastes and social norms as about the history of anatomical bodies. Just as public dissections in the Renaissance reflected and constructed contemporary norms regarding the body, the design, materialization, and dissemination of digital cadavers articulates our current standards with regard to the body and its transparency. Virtual anatomy thus reflects the anatomical theater of the twenty-first century, a theater in which education, morality, and entertainment are seamlessly woven into a digital culture. Our knowledge of the body can never be separated from its representation, the technology that mediates it, and the cultural matrix from which it arises.

EPILOGUE

Filmed operations, plastinated corpses, endoscopic travels through the body, X rays of lung tissue, ultrasound images of fetuses, digitized cadavers—all of these once challenged or still challenge our perception of the body. Together, they suggest an ingrained, historically changing desire for more access to, and more transparency of, the body's interior. Over the past centuries, physicians have increasingly explored the body beneath the skin, and, especially in the last hundred years, their insights have been spread through media such as film, photography, television, video, and the computer. As a consequence, the world of medicine is no longer confined to its scientific and professional domain, but has increasingly become interlaced with culture at large: doctor meets artist, medicine flirts with publicity, operating theater merges with living room. Because at the same time as the visual media have pervaded everyday life, the interplay of medicine and other domains such as art, media, ethics, and technology, has intensified and grown more complex and versatile. Cultural analysis offers one way to understand the mediation of bodies, both historical and contemporary.

When confronted with new or challenging representations of the body, our typical response is to pose concrete ethical questions: Should this be allowed? How far should we go in showing such images or applying such technologies? What is morally or ethically acceptable? In medicine, ethics are a matter of philosophical reflection resulting in professional agreements; the various ethical codes that result from debates within medical professional organizations are translated into binding protocols for their members. By contrast, in the media, ethics are commonly a matter of unwritten rules or informal agreements among broadcasting organizations—if not altogether left up to the individual documentary maker to decide the acceptability of what is shown.[1] Debates about media ethics rarely result in concrete standards for the profession as a whole, because there is no professional institution to impose sanctions. In media representations of medical issues, therefore, two normative systems frequently collide, each time triggering new questions about professional responsibilities, shared values, or individual taste.

In addition to questions of ethics, the issue of responsibility has been touched upon throughout this book. It is difficult in many cases to determine the various responsibilities at stake. Who, for instance, is responsible for the way in which operations are represented in documentaries? Who is responsible for the display or popularization of dead bodies in a museum or, as a digital data set, on the Internet? Of course, each profession has its own values and responsibilities, but the moment they converge in the public domain, a hierarchical relationship between the various interests emerges. In some cases, medical professionals are hardly aware of how the television camera depicts their activities, and, consciously or not, put all representational power into the hands of the camera crew. In other cases, television producers may uncritically adopt the angle of the *camera medica,* which subsequently determines what viewers will see. But the norms and values of medical and media professionals—or, for that matter, their tastes—may differ quite radically. It is not uncommon that what appears self-evident in anatomical laboratories or operation theaters takes on substantially different meanings outside these confined professional spaces. It is important, then, that medical professionals learn to anticipate the possible effects of medical representations. This same responsibility applies to artists and media producers.

Medical and media professionals have a shared responsibility in how patients are represented, especially patients without a voice of their own, such as fetuses, babies, seriously disabled persons, or the dead. Various scholars have pleaded for stricter guidelines regarding the participation of medical subjects in documentaries and television programs.[2] Even with respect to "ordinary" patients, restraint is called for. After all, patients are always vulnerable to some extent; they depend on the expertise of doctors, and a request to participate in some form of publicity may impose unnecessary or undesirable pressures on the patient. Patients, like doctors, can never be fully aware of the consequences of their consent to have operations broadcast on television. Yet it is equally true that patients, even though most have little or no training in analyzing medical representations, may be strongly attracted to seeing or having images of their bodies.[3] In the role of patient, viewer, or consumer, many people are happy to take home an ultrasound video of their unborn child, a filmed record of the endoscopic surgery of their knee, or some other visual medical souvenir.

The visual representation of the internal body, then, is the shared responsibility of medical doctors, media producers, patients, and viewers, rather than of a single individual or professional authority. There are no fixed rules or protocols for its regulation, so agreements on how to represent a case between, for instance, physicians and media producers, are always local, incidental, or provisional. My interest in tracing the intersections of medicine and media norms goes far beyond the incidental run-ins of doctors and television makers; in applying cultural analysis, the critic is more interested in the long-term transformations of cultural values, in the continuities and discontinuities that constitute the subtext of incidents and cultural products. This explains why public debates play a major role as an object of study, especially when certain representations of the body test the consensus. Public debates are crucial to our understanding of visual representation: they establish, at any given point in time, what the norms and values are, or, for that matter, divulge how they have changed. A critic of culture is not primarily engaged in defining the standards for what is allowed in representation and what is not. As public norms constantly evolve, it is this very dynamic that the cultural critic tries to map, analyze, and historicize.

Interconnected interests (both political and/or economic) also play a prominent role in the contact zone of media and medicine. In many television programs or documentaries, the evidence of shared interests is hard to miss. In the case of a plastic surgeon performing free surgery on the condition that the procedure and results will be broadcast as medical infotainment, collaboration is obviously meant to provide free public relations for the surgeon's private clinic (as, for instance, in programs like *Extreme Makeover*). Even when doctors or hospitals contribute to medical television programs because they feel the public might benefit from being informed about a new disease or the latest medical technology, their cooperation is, in fact, rarely without self-interest. This is not to suggest that something is wrong with having interests; it is crucial, however, that interests are identified as such, or that it is possible for outsiders to identify them. Medical technologies—expensive visualizing technologies in particular—basically sell themselves: the more we see of the human body, the more we want to see, as in the case of filmed operations and plastination. The case studies on ultrasound for fun and virtual cadavers demonstrate how new medical technologies gain prominence in the marketplace by their promotion in the public domain. Yet the economic or political interests for representing the body in a certain fashion are not always identifiable; there is reason for caution, especially when these motives are hidden or disguised.

When tracing the changing representations of the interior body in the public domain—in television programs, advertising, stories, art, or on the Internet—concrete opportunities for linking the various interests come into play. One major concern of cultural analysis is to disentangle these linked interests and sort them out. As in the case of responsibility for representation, the cultural critic is not primarily interested in playing the role of arbitrator. It is relatively easy to dismiss sensational displays of the body's interior but much more important to analyze how these displays are rooted in cultural and historical contexts. It also takes little effort to embrace or denounce the increasing interconnection of advertising, information, knowledge, and entertainment; showing how, exactly, this process is occurring and evolving is an altogether different matter. Mapping and understanding normative transformations in culture at large comes prior to articulating value judgments about these changes.

In addition to addressing questions about responsibility for images and the interconnection of interests, this study poses questions associated with transformations in our cultural norms and values concerning the human body. What does a genre, television program, or work of art reveal about our standards regarding bodily perfection, beauty, and health? Have these standards changed over time, and if so, how is this expressed? How do views of privacy and bodily integrity change on account of newly developed visualization technologies? What is the role of gender and ethnicity in the construction of cultural norms and values, especially regarding bodily (im)perfection? How has the relationship between media and medicine, or scientists and artists, changed over time? And, at a more general level, what is the effect of large cultural processes, such as mediation and the obfuscation of genre boundaries, on concrete practices and contexts of human-body representation? Most of these questions do not have simple answers; on the contrary, most of the answers trigger new, more pertinent questions. In addressing these questions, I have tried to understand in what ways bodies really are mediated, and how the processes of mediation render the body a constantly changing locus of culture.

A cultural analysis of medical imaging in everyday life should move beyond merely evaluating individual or incidental cases. Instead, it is aimed at historicizing and contextualizing individual phenomena and expressions, in an attempt to discover, establish, trace, and understand how they fit larger cultural patterns. This kind of scholarly analysis offers a broad basis for critique, reflection, appreciation, and insight into our culture. As this book has shown, popular medical television programs are rooted in a cinematic tradition of medical films. The recently invented plastination technique is directly linked to a long-standing tradition in anatomy and art; its application not only capitalizes on the mystery of the dead body (which, as the success of the Bodyworlds exhibition suggests, continues to fascinate us) but adds a new chapter to that history. Representations of the body that are made possible by new tools and technologies reveal new dimensions of the body, consequently changing it. A century ago the endoscope triggered the fantasy of journeying through the body; the most recent virtual technologies seem to have made such a journey a reality. At the same time, this accomplishment will change our standards regarding health and bodily perfection, and in turn give rise to new dilemmas about medical intervention. At the start of the twentieth century, the invention of X rays dramatically changed our conception of the internal body; at the close of that same century it became clear that ultrasound not only has immediate medical significance in pregnancies, but also effectuates major psychological and cultural meanings. Studying the interplay of media and medical technologies, and situating this interplay historically, demonstrates that the body, in addition to being a medical construct, is always also a cultural construct. Transparency is not merely an ideal nurtured in medicine, nor is it exclusively a fantasy cultivated by media and advertising. Instead, the ideal of a body that is malleable, perfectible, transparent, and fully understood can be traced at many levels in our culture.

By starting from concrete cases and exploring their coherency and historicity, the contours of the ongoing interplay of medicine, technology, media, and culture come into view. The developments in medical imaging illustrated here cannot be isolated from more general developments in Western culture, which has, especially in the past century, transformed itself into a culture of images.[4] The great value attached to visual data in medicine as proof of an organism's health or disease is closely intertwined with the growing contribution of photos and films to the construction of ethical, aesthetic, and cultural norms. This interrelationship is intricate and multifaceted; its analysis, rarely straightforward or unambiguous. The ideal of transparency is constantly evolving as we go along conceptualizing and developing the very tools for corporeal elucidation.

NOTES

CHAPTER 1

MEDIATED BODIES AND THE IDEAL OF TRANSPARENCY

1 For a theoretical and historical overview of the intersections between medicine and culture, see, for instance, Deborah Lupton, *Medicine as Culture: Illness, Disease and the Body in Western Societies* (London: Sage, 1994).

2 Stanley J. Reiser, "Technology and the Use of the Senses in Twentieth-Century Medicine," in *Medicine and the Five Senses,* ed. Catherine Byrum and Roy Porter (Cambridge: Cambridge University Press, 1996), 262–73, provides an illuminating introduction to the nineteenth-century transformation from sensory diagnostics to objective mechanical diagnostics.

3 For an eminent and insightful analysis of the work of the French engineer Etienne-Jules Marey, who is also acknowledged as the "godfather of film," see Marta Braun, *Picturing Time: The Work of Etienne-Jules Marey, 1830–1904* (Chicago: Chicago University Press, 1992).

4 This short list of techniques is far from exhaustive. There are imaging techniques that are basically variations or specific applications of X ray, such as mammography, and instruments based on a combination of various techniques, such as fluoroscopy—a combination of X-ray technology, image intensifier, and television-screen technology. For a technical, yet lucid, and for laypersons understandable, description of medical imaging technologies, see Anthony B. Wolbarst, *Looking Within: How X-Ray, C.T., M.R.I., Ultrasound, and Other Medical Images Are Created and How They Help Physicians Save Lives* (Berkeley: University of California Press, 1999).

5 See, for instance, T. Doby and G. Alker, *Origins and Development of Medical Imaging* (Carbondale, Southern Illinois Press, 1997), who argue: "Having reviewed the phases of how we accumulate knowledge by seeing, we realize that the speed of advance has increased geometrically. . . . We are in the midst of a great leap. Science does not grow in a continuum but—as quanta of energy—jumps in packages" (131). See also Wolbarst, *Looking Within,* chapter 4, "Shadows in Computers: Going Digital."

6 For an example of a historical study of particular medical imaging instruments, see, for instance, Stuart Blume, *Insight*

and Industry: On the Dynamics of Technological Change in Medicine (Cambridge, Mass.: MIT Press, 1992).

7 Bettyann Holtzmann-Kevles, *Naked to the Bone: Medical Imaging in the Twentieth Century* (New Brunswick: Rutgers University Press, 1997), 3.

8 Holtzmann-Kevles, *Naked to the Bone,* 3.

9 For a detailed elaboration of this theory on the mutual shaping of bodies, machines, and procedures, see Bernike Pasveer, "Knowledge of Shadows: The Introduction of X-Ray Images in Medicine," *Sociology of Health and Illness* 11 (1989): 360–81.

10 According to Barron H. Lerner, "The Perils of X-Ray Vision: How Radiographic Images Have Historically Influenced Perception," *Perspectives in Biology and Medicine* 35 (1992): 382–97, it was several decades before the interpretation of shadows on X rays, as a diagnosis of tuberculosis, stabilized.

11 The imaging sciences, as Don Ihde suggests in "Virtual Bodies," in *Body and Flesh: A Philosophical Reader,* ed. Don Welton (London: Blackwell, 1998), 349–57, have contributed to the emergence of an intricate hermeneutical visual praxis.

12 See; for instance, Steve Woolgar and Michael Lynch, eds., *Representation in Scientific Practice* (Boston: MIT Press, 1990); Don Ihde, *Expanding Hermeneutics: Visualization in Science* (Chicago: Northwestern University Press, 1999); and Bruno Latour, "Visualization and Cognition: Thinking with Eyes and Hands," in *Knowledge and Society: Studies of the Sociology of Culture,* ed. Henrika Kuklich and Elizabeth Lang, (London: Jai Press, 1986), 1–40.

13 Ian Hacking, *Representing and Intervening: Introductory Topics in the Philosophy of Natural Science* (Cambridge: Cambridge University Press, 1995), 192.

14 For a discussion on the validation of the Gulf War syndrome and chronic fatigue syndrome (CFS) by medical imaging techniques, see Joseph Dumit, "When Explanations Rest: 'Good-enough' Brain Science and the New Socio-medical Disorders," in *Living and Working with the New Medical Technologies: Intersections of Inquiry,* ed. Margaret Lock, Allan Young, and Alberto Cambriosio (Cambridge: Cambridge University Press, 2000), 209–32.

15 In using the term "cultural analysis" rather than the broader term "cultural studies," I follow Mieke Bal, who argues that cultural analysis is only meaningful if deployed in close interaction with the objects of study to which it pertains. See Mieke Bal, ed., *The Practice of Cultural Analysis: Exposing Interdisciplinary Interpretation* (Stanford: Stanford University Press, 1999).

16 This estimate is made by Wolbarst, *Looking Within,* xi.

17 These numbers are quoted from a report by the Blue Cross and Blue Shield Association, in Reed Abelson, "An M.R.I. Machine for Every Doctor? Someone Has to Pay," *New York Times,* March 13, 2004, Business section.

18 For an extensive analysis of the connection between X rays and cinema, see chapter 5 of Lisa Cartwright, *Screening the Body: Tracing Medicine's Visual Culture* (Minneapolis: University of Minnesota Press, 1995).

19 The term "mediation of culture" is theorized extensively by John B. Thompson, *The Media and Modernity: A Social Theory of the Media* (Cambridge: Polity Press, 1995), who defines it as a historical, ongoing series of transformations

structured by modes of human communications and their institutions; the latest stage in this series of transformations is what Thompson describes as a world of "mediated experiences."

20 This event, recorded and displayed through a webcam, happened on June 14, 2000, and was broadcast live on the Web site of a popular woman's magazine.

21 Kim Sawchuck, "Enlightened Visions, Somatic Spaces: Imaging the Interior in Art and Medicine," in *RX: Taking Our Medicine* (Exhibition Catalogue, Toronto, 1999), no page numbers.

22 Michel Foucault, *The Birth of the Clinic: An Archeaology of Medical Perception* (Londen: Travistock, 1973), xiv.

23 Jackie Stacey, *Teratologies: A Cultural Analysis of Cancer* (London: Routledge, 1998), provides an impressive cultural analysis of her own struggle against a rare form of ovarian cancer; she convincingly argues that the processes of medical diagnosing are closely related to cultural connotations and the proposed treatment of disease. Although her book is much broader in scope, she specifically acknowledges the role of medical imaging technologies in this process: "The significant knowledge about what is going on inside is captured as external image or code, mediated through technological processes which have invisible, though often damageable effects" (157–58).

24 For an extensive description of the social and cultural construction of in vitro fertilization and infertility, see José van Dijck, *Manufacturing Babies and Public Consent: Debating the New Reproductive Technologies* (New York: New York University Press, 1995); see also Sarah Franklin, "Postmodern Procreation: Representing Reproductive Practice," *Science as Culture* 3 (1993): 522–61.

25 For an interesting analysis of the representation of viruses in popular culture, see Sarah Jain, "Mysterious Delicacies and Ambiguous Agents: Lennart Nilsson in *National Geographic*," *Configurations* 6 (1998): 373–95.

26 See, for instance, Scott Montgomery, *The Scientific Voice* (New York: The Guilford Press, 1996), particularly chapter 3, "Illness and Image."

27 For an analysis of metaphorical images and fictional narratives on genetics, see José van Dijck, *ImagEnation: Popular Images of Genetics* (New York: New York University Press, 1998).

28 The ubiquitous presence of medical images in mass media should prompt better education of people's "visual literacy," argues Jane Trumbo, "Visual Literacy and Science Communication," *Science Communication* 20 (1999): 409–25.

29 T. Hugh Crawford, "Visual Knowledge in Medicine and Popular Film," *Literature and Medicine* 17 (1998): 24–44. Quotation on page 42.

30 Friedrich A. Kittler, *Discourse Networks, 1800–1900,* trans. Michael Metter (Stanford: Stanford University Press, 1990); originally published in German as *Aufschreibesysteme 1800–1900* (München: Wilhelm Fink Verlag, 1985). Kittler describes the transition to inscription technologies at the end of the nineteenth century, particularly in part 2.

31 I am referring particularly to Lisa Cartwright's analysis of X-ray images as cultural technologies in *Screening the Body,* and to Jackie Stacey's analysis of her understanding of visual representations of cancer in her *Teratologies.*

32 See, for instance, Joseph Dumit, "Brain-Mind Machines and American Technological Dream Marketing," in *Cyborgs and Citadels: Anthropological Interventions in Emerging Sciences and Technologies,* ed. Gary L. Downey and Joseph Dumit (Santa Fe: School of American Research Press, 1999), 347–62; and Joseph Dumit, "Objective Brains, Prejudicial Images," *Science in Context* 12 (1999): 173–201. Dumit contends that the development of PET scans, far from being a pure technological innovation, involves a highly complex process with a variety of disciplinary inputs. See also Anne Beaulieu, "The Brain at the End of the Rainbow: The Promises of Brain Scans in the Research Field and in the Media," in *Wild Science: Reading Feminism, Medicine and the Media,* ed. Janine Marchessault and Kim Sawchuk (New York: Routledge, 2000), 39–54. Anthropologist Emily Martin's analysis of microscopic images of the HIV virus shows that they have played a significant role in both medical and public definitions of AIDS; see her *Flexible Bodies: The Role of Immunity in American Culture from the Days of Polio to the Age of AIDS* (Boston: Beacon Press, 1994).

CHAPTER 2
THE OPERATION FILM AS A MEDIATED FREAK SHOW

1 For a detailed historical description of freaks and freak shows, see Leslie Fiedler, *Freaks: Myths and Images of the Secret Self* (New York: Simon and Schuster, 1978).

2 Elizabeth Grosz, "Intolerable Ambiguity: Freaks as/at the Limit," in *Freakery: Cultural Spectacles of the Extraordinary Body,* ed. Rosemarie G. Thomson (New York: New York University Press, 1994), 55–66. Quotation on p. 56.

3 In 1573, Ambroise Paré published his famous tract "Des Monstres et Prodiges," in which he classified "monstrous people" on the basis of pseudoscientific terminology.

4 Michel Foucault, *The Birth of the Clinic: An Archeaology of Medical Perception* (London: Travistock, 1973).

5 George Gould and Walter Pyles, in their medical encyclopedia *Anomalies and Curiosities of Medicine* (New York: W. B. Saunders, 1896), refrained from every religious or mythical explanation.

6 The American Museum, established by P. T. Barnum in 1841, may be the most famous example of a freak show with a stable basis in New York. The Barnum and Bailey and Ringling Brothers circuses, which traveled around the United States and Europe, contracted "freaks" to join their show for a season. For an elaborate history of the nineteenth-century American freak show, see Robert Bogdan, *Freak Show: Presenting Human Oddities for Amusement and Profit* (Chicago: University of Chicago Press, 1988).

7 Rosemarie G. Thomson, in her introduction to *Freakery,* argues that freaks were always exploited: "Like the bodies of females and slaves, the monstrous bodies exist in societies to be exploited for someone else's purposes" (2).

8 For an elaborate analysis of racial and ethnic elements of the nineteenth-century freak show, see Ben Lindfors, "Ethnological Show Business: Footlighting the Dark Continent," in Thomson, *Freakery,* 207–18; see also Christopher A. Vaughan, "Ogling Igorots: The Politics and Commerce of Exhibiting Cultural Otherness," in Thomson, *Freakery,* 219–33.

9 Ronald E. Ostman, "Photography and Persuasion: Farm Security Administration Photographs of Circus and Carnival Sideshows 1935–42," in Thomson, *Freakery,* 121–36, explains how pictures of circus freaks concurrently served as public relations tools and as verification instruments for their scientific authentication.

10 Katherine Park and Lorraine Daston, in their elaborate study of the "wonders of nature," have illustrated how, since the Renaissance, emphasis on the public display of deviant bodies shifted from monsters as prodigies to monsters as examples of medical pathology. See Katherine Park and Lorraine Daston, "Unnatural Conceptions: The Study of Monsters in Sixteenth and Seventeenth Century France and England," *Past and Present* 92 (1981): 23–46.

11 Several biographies have been written on the lives of Eng and Chang Bunker, and their American and European performances have been documented in hundreds of newspaper articles and pamphlets. For this chapter, I have consulted Irving Wallace and Amy Wallace, *The Two* (New York: Simon and Schuster, 1978).

12 Hillel Schwarz, *The Culture of the Copy: Striking Likeness, Unreasonable Facsimiles* (New York, Zone Books, 1996), relates the life story of the Bunker twins to the general tragedy of slavery in nineteenth-century America; she argues that the case of Chang-Eng reflects the larger dilemmas of slavery and otherness in American postabolition society (52–59). Allison Pingree, "America's 'United Siamese Brothers': Chang and Eng and Nineteenth-Century Ideologies of Democracy and Domesticity," in *Monster Theory: Reading Culture,* ed. Jeffrey J. Cohen (Minneapolis: University of Minnesota Press, 1996), suggests that the Bunker twins, at the time of the Civil War, were both a symbol of national unity and a symbol of dubious moral standards because their conjoined bodies and their marriage to the Yates sisters gave rise to disputes about group sex, homoeroticism, incest, and adultery.

13 David Gerber, "The 'Careers' of People Exhibited in Freak Shows: The Problem of Volition and Valorization," in Thomson, *Freakery,* 38–54, criticizes Bogdan's explanation of the gradual disappearance of the freak show in the early decades of the twentieth century. According to Gerber, the reason for its decreasing popularity was a general change in mentality; people began to feel social embarrassment at the public exploitation of handicapped people. I agree with Gerber that the disappearance of the freak show requires a more complex explanation than the one Bogdan provides. Such discussion, however, is beyond the scope of this book.

14 Cinema incorporated many elements that were already an integral part of contemporary mass entertainment. Vanessa Schwartz, "Cinematic Spectatorship before the Apparatus: The Public Taste for Reality in Fin-de-Siècle Paris," *Viewing Positions: Ways of Seeing Film,* ed. Linda Williams (New Brunswick, N.J.: Rutgers University Press, 1995), 87–113, argues how precinematic audiences in the nineteenth century already showed a remarkable preference for a combination of the "authentic" and the extraordinary. See also Tom Gunning, "The Cinema of Attraction: Early Film, Its Spectator, and the Avant-garde" *Wide Angle* 8 (1986): 63–70, who argues that early cinema has its roots in public spectacles.

15 The distribution of Tod Browning's film *Freaks* (1932), an MGM production, was canceled after public outrage over the film's blatant and aggressive depiction of handicapped persons. For a historical and contextual analysis of this film, see Joan Hawkins, "One of Us: Tod Browning's *Freaks*," in Thomson, *Freakery*, 265–76.

16 Erin O'Connor, "Camera Medica: Towards a Morbid History of Photography," *History of Photography* 23 (1999): 232–44, describes the hybrid genre of the medical photograph as follows: "Part portrait, part record, medical photographs were, to say the least, confusing things to behold. As portraits, they were clearly artistic (many were even taken in the studios of professional photographers). But as portraits of diseased, deformed, and deranged individuals, they were frequently far from aesthetic, and occasionally hard to identify as pictures of people" (232).

17 The United States pioneered the genre of medical documentaries on television. As early as 1951, an American station broadcast the recording of a heart surgery. In the Netherlands, the distribution of operation movies remained virtually restricted to the medical circuit until the 1960s. For details on the history of the American medical film, see Frank Warren, *Television in Medical Education Handbook* (Washington, D.C.: American Medical Association, 1955); see also Michael Essex-Lopresti, "Centenary of the Medical Film," *Lancet* 349 (1997): 819–20; and Michael Essex-Lopresti, "The Medical Film, 1897–1997: Part I," *Journal of Audiovisual Media in Medicine* 21 (1998): 48–55.

18 Guy Debord, *The Society of the Spectacle* (Detroit: Black and Red, 1977).

19 According to Claude Edelmann, "A la Découverte du Corps Humain," in *Le Cinéma et la Science,* ed. A. Martinet (Paris: CNRS, 1994), 174–81, narrative is one of the most defining features of twentieth-century medical and science documentaries, as it obeys the first law of cinema: to tell a good story.

20 For an introduction to Dr. Doyen's films, see *Cinéma et Médecine* a special edition on the occasion of the centennial of the film, distributed by the Musee d'Histoire de la Médecine in Paris, 1996.

21 See Thierry Lefèbvre, "Le docteur Doyen, un précurseur" in Martinet, *Le Cinéma et la Science,* 70–77; and Jean-Michel Arnold, "La Grammaire Cinématographique: Une Invention des Scientifiques," in Martinet, *Le Cinéma et la Science,* 211–17.

22 Thierry Lefèbvre, "La Collection," *1895* 17 (1994): 100–114.

23 The original copy of *Sepération des Soeurs Siamoises Xiphopages Doodica et Radica* has been lost. A short fragment of several minutes is archived in the Institut de la Cinématographie, Paris.

24 The commentary Doyen gave in his lecture has been published as "Le Cas des Xiphopages Hindoues Radica-Doodica," *Revue Critique de Médecine et de Chirurgie* 4 (1902): 3031. Cited in *Cinéma et Médecine,* 16.

25 For a detailed description of the lawsuits Doyen initiated and the protracted legal skirmish following Doyen's discovery, see Lefèbvre, "La Collection."

26 The oral pleadings of Monsieur Desjardin, Doyen's lawyer, are interesting for more than one reason, but primarily because he argues convincingly for an exclusive medical reading of Doyen's films, allowing him to lash out at the "opérateurs" who had defamed the work

of a decent surgeon. See the court document *Contrefaçon Artistique des Cinématographies du Docteur Doyen: Plaidoirie de Me Desjardin, Avocat à la Cour d'Appèl,* Jugement du Tribunal (10 Février 1905); original deposited in the George Eastman Archive, Rochester, New York.

27 According to Doyen's strict terms, the operation films could only serve "pour servir l'enseignement de la technique opératoire et de la demonstrations des progrès realistes par les grandes opérations chirurgicales." Cited in Lefèbvre, "La Collection," 110.

28 I found approximately ten original copies of operation films at the archive of the Dutch Film Museum in Amsterdam; most (silent) films are stored on videotapes and are between two and fifteen minutes in length. Shots of the actual operation are typically alternated with short textual explanations.

29 The Dutch Scientific Film Association (NVWF) consisted of five sections, but the medical section was by far the most active. Their meeting minutes, program schedules, and general brochures are stored at the National Audiovisual Archive, previously called the Archive of Film and Science, Amsterdam, Netherlands.

30 *The Frisian Conjoined Twins,* (UNFI), 44 minutes, color, premiered on October 16, 1954, at the general meeting of the Dutch Medical Association (KNMG). The archived records of the NVWF prove that this film was widely distributed in Europe and the United States. The Dutch film on the separation of conjoined twins is not the only one of its genre; several months before this Dutch production, a similar operation was filmed in London. According to the academic report of this procedure, the Dutch and British films are quite similar in terms of film techniques and narrative format. For more information on the British film, see Stanley Schofield and Michael Essex-Lopresti, "Filming the Surgical Separation of the Conjoined Twins of Kano" *Science and Film* 3 (1954): 19–23. The *American Medical and Surgical Motion Pictures Catalogue* (Chicago: AMA, 1964) lists five filmed operations on conjoined twins, one of which is a copy of the Dutch movie discussed here.

31 An article written by the producers of this film, W. de Vogel, J. W. Varossieau, and R. Blumenthal-Rothschild, "The Film of the Frisian Conjoined Twin," *Science and Film* 3 (1954): 20–25, is quite candid about its intentions: "The film should not only be a record of a rare operation (of which the value to medical students is very doubtful) but should also deal with the whole subject of double monstra leading up to the present case, which would include parturition, preliminary investigations and the operation" (21).

32 The minutes of the Dutch Scientific Film Association (NVWF) reflect the concern of several members about an increasing dramatization on and commercialization of the medical film. In 1960, one of the board members complained that popular and dramatized scientific films begin to outnumber "serious scientific productions" at the yearly conferences of the International Scientific Film Association: "The ISFA should provide an exact definition of 'scientific film,' which would effectively eliminate the overload of popular science films at our meetings." In 1961, another board member objected to an

American operation movie shown at one of the association's screenings, which was interrupted by a commercial for Upjohn, a pharmaceutical company.

33 Since 1989, a Dutch public station has broadcast the program *Surgeon's Work (Chirurgenwerk),* a hybrid talk show–documentary featuring a different operation each week (see also chapter 4 of this book). In the United States, *The Operation* used to be a weekly broadcast on the Learning Channel, produced by Advanced Medical Education.

34 Catherine Belling, "Reading *The Operation:* Television, Realism, and the Possession of Medical Knowledge," *Literature and Medicine* 17 (1998): 1–23. Citation on page 1.

35 *Siamese Twins,* directed and produced by Jon Palfreman, broadcast on WGBH Boston and BBC Television in 1995, as part of the series NOVA/Horizon, a coproduction of PBS (USA) and BBC (Great Britain).

36 David Clark and Catherine Myser, "Being Humaned: Medical Documentaries and the Hyperrealization of Conjoined Twins," in Thomson, *Freakery,* 338–55, provide an insightful analysis of Palfreman's *Siamese Twins.* One of their most pointed criticisms is that the separation of conjoined twins is presented as a sine qua non for an acceptable form of life. Whether these twins, as a rule, should be separated at all is never raised as a legitimate question, and the surgical separation of shared organs on the basis of strictly medical criteria, according to Clark and Myser, is "impurely utilitarian in nature" (346).

37 Despite the growing presence of imaging and diagnostic techniques, conjoined twins are still born in the United States and other Western nations. The medical imperative of surgical separation, various critics have argued, is so dominant that parents of conjoined twins hardly ever dare question the inevitability of the operation. For a careful analysis of the ethical dilemmas involving the surgical imperative, see David Thomasma, "The Ethics of Caring for Conjoined Twins: the Lakeberg Twins" *The Hastings Center Report* 26 (1992): 4–12; see also Catherine Myser and David Clark, "Fixing Katie and Eilish: Medical Documentaries and the Subjection of Conjoined Twins," *Literature and Medicine* 17 (1998): 45–67.

38 Clark and Myser, "Being Humaned," 343.

39 *Conjoined Twins* (Robert Eagle, prod., James van der Pool, dir.), a Horizon documentary, was broadcast on October 19, 2000. The title of this documentary, in contrast to the one by Palfreman, is already indicative of the different ideological approach to this subject.

CHAPTER 3
BODYWORLDS: THE ART OF PLASTINATED CADAVERS

1 In the exhibition's catalogue, *Körperwelten: Einblicke in den menschlichen Körper* (Heidelberg: Institut für Plastination, 1997), von Hagens defines anatomical art as the "ästhetisch instruktive Darstellung des Körperinneren" (217).

2 Part of that discussion, a public dialogue to which medical scientists, theologians, art historians, lawyers, and journalists contributed their point of view, is reflected in the collection of essays edited by Franz J. Wetz and Brigitte Tag, *Schöne Neue Körperwelten: Der Streit um die Ausstellung* (Stuttgart: Klett-Cotta, 2001).

3 For a general introduction into the his-

tory of anatomy and anatomical models, see Roy Porter, *The Greatest Benefit to Mankind: A Medical History of Humanity from Antiquity to the Present* (London: Harper Collins, 1997).

4 Christina Lammer, *Die Puppe: Eine Anatomie des Blicks* (Vienna: Verlag Turia, 1999), provides an insightful philosophical analysis of the use of body models or "puppets," as she calls the various anatomical objects produced between the early sixteenth and late twentieth century.

5 Katherine Park, "The Criminal and the Saintly Body: Autopsy and Dissection in Renaissance Italy," *Renaissance Quarterly* 1 (1994): 1–33, argues that opening the body and later embalming or preserving it, either in its entirety or in parts, was a common funerary practice as early as the twelfth century. In Italy, the corpses of candidates for sainthood were dissected to examine them for miraculous marks, and were subsequently eviscerated. Between the twelfth and the early sixteenth centuries, various primitive techniques were used to preserve corpses, from simply boiling and drying the anatomical object to tanning it by soaking it in honey or wine. Anatomists like Ambroise Paré (1510–1590) experimented with the usefulness of alcohol in embalming. On these practices, see, for instance, F. Gonzalez-Crussi, *Suspended Animation: Six Essays on the Preservation of Bodily Parts* (San Diego: Harcourt Brace, 1995).

6 Julie V. Hansen, "Resurrecting Death: Anatomical Art in the Cabinet of Dr. Frederick Ruysch," *Art Bulletin* 78 (1998): 663–79. Citation on page 664.

7 Hansen, "Resurrecting Death," 676.

8 Ludmilla Jordanova, "Medicine and the Genres of Display," in *Visual Display: Culture Beyond Appearances,* ed. Lynne Cooke and Peter Wollen (Seattle, Bay Press, 1995), 202–17.

9 For a general overview of the laws and restrictions regulating the dissection of corpses, see chapter 2 of Ruth Richardson, *Death, Dissection and the Destitute* (London: Routledge, 1987).

10 Wax models, for instance, are preferable to real bodies when it comes to showing diseases of the skin and specific dermatological pathologies. See Thomas Schnalke, *Disease in Wax: The History of Medical Moulage* (London: Quintessence, 1995).

11 For a description of the Italian school of wax modeling in the seventeenth and eighteenth centuries, see chapter 1 in Barbara Stafford, *Body Criticism: Imaging the Unseen in Enlightenment Art and Medicine* (Cambridge, Mass.: MIT Press, 1993). A short overview of the Bolognese collections can be found in Karen Newman, *Fetal Positions: Individualism, Science, Visuality* (Standford: Stanford University Press, 1996).

12 A large collection of wax models, primarily produced by Clemente Sussini and Paolo Mascagni, is still exhibited in La Specola in Florence; for further reading on the history of these models, see Rummy Hillowala, *The Anatomical Waxes of La Specola* (Florence: Arnaud, 1995). The Pathologisches-Anatomisches Museum Josephineum in Vienna also harbors quite a few samples. For a history of exhibited wax models, see Gabriela Schmidt, "Sammlung Anatomischer und Geburtshilflicher Wachsmodelle," in *Catalogue Institut für Geschichte der Medizin der Universität Wien* (Vienna, 1999), 37–40.

13 Wax models, in spite of their beauty and

accuracy, had one important drawback: they were very vulnerable. For that very reason, Dr. Louis Thomas Auzoux (1797–1880) experimented with papier-mâché and gradually developed body models that were both accurate and "touch-resistant." His models became very popular in teaching hospitals, schools, academies of art, and veterinary schools.

14 Lorraine Daston and Katherine Park, *Wonders and the Order of Nature, 1150–1170* (New York: Zone Press, 1998), describe the long and diverse tradition of the "wonders of nature" and the "wonders of art," from medieval tradition to the cabinets of curiosities or *(Kunst-* and *Wunderkammer)* in the Renaissance. They illustrate how, within this tradition, astonishment and horror, the sublime and terror, often coincided in the collections on display. See especially chapter 7, "Wonders of Art, Wonders of Nature," 255–301.

15 *Körperwelten,* 195–207.

16 In the United States, Dr. Roy Glover, professor of anatomy at the University of Michigan, in collaboration with Dow Corning, has slightly modified von Hagens's plastination technique, which he applies to the preparation of complete bodies as well as body parts. After replacing the body fluids with a silicone polymer that contains a catalyst, the body hardens within twenty-four hours. According to the *New York Times,* Glover plastinates bodies only for educational purposes, and some of his cadavers will soon be on display in a traveling anatomical museum. See Mary Roach, "A New Student Aid: Plastic Body Parts, Made from the Real Things," *New York Times* March 7, 2000.

17 Donna Haraway, *Simians, Cyborgs, and Women: The Reinvention of Nature* (New York, Routledge, 1991).

18 For various interesting angles on this issue, see, for instance, the collection of essays edited by George Robertson and others, *Future Natural: Nature, Science, Culture* (London: Routledge, 1996). See also Nelly Oudshoorn, *Beyond the Natural Body: An Archaeology of Sex Hormones* (London: Routledge, 1994).

19 Erwin Panofsky, "Artist, Scientist, Genius: Notes on the 'Renaissance-Dämmerung,'" in *The Renaissance: Six Essays,* ed. Wallace K. Ferguson (New York: Academy Library, 1953), 123–82. A clear interpretation of Panofsky's views in relation to anatomical aesthetic conventions is provided by Glenn Harcourt, "Andreas Vesalius and the Anatomy of Antique Sculpture," *Representations* 17 (1987): 28–61.

20 A hierarchical tension typically impacted working relationships between anatomists and artists, who commonly teamed up to wed scientific precision and artistic refinement in the production of anatomical atlases. These professional marriages of opposite disciplinary bedfellows did not always go smoothly. It is a well-known fact, for instance, that the Dutch anatomist Bernhard Siegfried Albinus and his artist-assistant Jan Wandelaar were engaged in constant fights over the scientific accuracy of Wandelaar's illustrations, as Albinus attempted to police his assistant's artistic judgment. See Catalogue Museum Boerhaave, *De Volmaakte Mens: De Anatomische Atlas van Albinus en Wandelaar* (Leiden: Boerhaave Museum, 1991).

21 See Harcourt, "Andreas Vesalius," 33–36.

22 On the symbolic and functional properties of body parts, particularly the hand, see Katherine Rowe, "God's Handy

Worke," in *The Body in Parts: Fantasies of Corporeality in Early Modern Europe,* ed. David Hillman and Carla Mazzio (London: Routledge, 1997), 285–312.

23 For a number of historical perspectives on the relationship between anatomical representations and artistic conventions, see various essays in Kathleen Adler and Marcia Pointon, eds., *The Body Imaged: The Human Form and Visual Culture since the Renaissance* (Cambridge: Cambridge University Press, 1993).

24 Jordanova, "Medicine and Genres of Display," 210.

25 Lorraine Daston and Peter Galison, "The Image of Objectivity," *Representations* 40 (1992): 81–128.

26 Erin O'Connor, "Camera Medica: Towards a Morbid History of Photography," *History of Photography* 23 (1999): 232–44, superbly explores the tension between mechanical objectivity and artistic subjectivity in relation to medical photography in the mid-nineteenth century, concluding: "For once we understand the time-honored distinction between scientific photography's documentary operations and art photography's aesthetic pretensions as a rationalizing fiction rather than as a statement of fact" (243). Barron H. Lerner has extensively discussed the problem of "mechanical objectivity" in relation to X rays in the nineteenth and early twentieth century in "The Perils of X-ray Vision: How Radiographic Images Have Historically Influenced Perception," *Perspectives in Biology and Medicine* 35 (1992): 382–97.

27 Daston and Galison, "Image of Objectivity," 98.

28 In her wonderful analysis and description of the collaboration between Joel-Peter Witkin (photographer) and Dr. Stanley Burns (ophthalmologist and collector of historical medical photographers), Rachelle A. Dermer refutes the assumption that the body is "a display of readable pathology, and that the photograph is the objective recorder," forcing viewers to interpret the body pathologically "as if they were interpreting that body and not the photograph" (246). As von Hagens did with his plastinated sculptures, Witkin and Burns collaboratively produced cultural objects that combined the authorities of photography and medicine to authenticate the human body as an object of meaning. See Rachelle A. Dermer, "Joel-Peter Witkin and Dr. Stanley B. Burns: A Language of Body Parts," *History of Photography* 23 (1999): 245–53.

29 For an informative and beautifully illustrated overview of the history of anatomical illustration, see K. B. Roberts and J. D. W. Tomlinson, *The Fabric of the Body: European Traditions of Anatomical Illustration* (Oxford: Clarendon Press, 1992); and K. B. Roberts, *Maps of the Body: Anatomical Illustration through Five Centuries* (St. John's: Memorial University of Newfoundland Press, 1981).

30 Even Vesalius was not very original: his muscle man carrying his own skin was a wink to the painting of the holy Bartholomew by Michelangelo, whose (self-)portrayal is featured on the ceiling of the Sistine Chapel (see figure 11).

31 The eclectic combination of artistic and anatomical modes of representation indicates that there has been an ongoing dialogue between various stylistic traditions and techniques of preservation across the centuries. The art of plastination both legitimizes and parodies such styles and traditions; this constant alter-

nation between imitation and parody is typical of a postmodern sensibility. For an introduction on the use of historical styles, pastiche, and parody in postmodernist art and literature, see Linda Hutcheon, *The Politics of Postmodernism* (London: Routledge, 1989), or Tim Woods, *Beginning Postmodernism* (Manchester: Manchester University Press, 1999).

32 Hillel Schwarz, *The Culture of the Copy: Striking Likeness, Unreasonable Facsimiles* (New York: Zone Books, 1996), addresses the dilemmas of authenticity, duplicity, and originality in a rather eclectic way, encompassing everything from Xerox machines to self-portraits and Siamese twins. A basic assumption is that the original sin is no less original in its numerous reenactments. Although Schwarz convincingly shows that the "culture of the copy" is nowhere near an exclusive feature of the postmodern era, the ethicality of the authenticity issue resurfaces most powerfully in the context of postmodern genres.

33 Representations of corporeal cross sections can be found even in the Middle Ages—for instance, in drawings of decapitated martyrs and saints. See Gonzales-Crussi, *Suspended Animation,* chapter 2.

34 *Körperwelten,* 214.

35 The plastinated sculptures challenge existing visual conventions as well as the act of observation. In his landmark study, *Techniques of the Observer: On Vision and Modernity in the Nineteenth Century* (Cambridge, Mass.: MIT Press, 1990), Jonathan Crary introduced the term "modernizing vision" to refer to the interdependence of new modes of looking (a flexible, physiologized gaze), instruments of visualization (camera obscura, and later the movie camera), and social and cultural developments in the mid-nineteenth century (mobility). Perhaps it is defensible to extend Crary's theory to the twentieth century, and regard MRI and CT techniques as technological tokens for "postmodernizing vision." This technological innovation not only enabled the observer to adopt an infinite variety of vantage points, but also prompted her to switch between different visual registers, both two-dimensional and three-dimensional. The emergence of this new way of looking cannot be isolated from a society in which the gaze has become de-physiologized, meaning that digital reconstructions of realities allow the viewer to take multiple angles and positions, even those which they can physiologically never adopt.

36 On the emergence and disappearance of anatomical theaters, especially in Italy, see Giovanni Ferrari, "Public Anatomy Lessons and the Carnival: The Anatomy Theatre of Bologna," *Past and Present* 117 (1987): 50–106.

37 Luke Wilson shows how spectators originally caught sight of the inside of the body while being confronted by violence and pain. Gradually, the notion of the reconstitution of the body began to dominate the anatomical dissection, as it yielded to a more clinical gaze in which not the dissection but the formation of a "body of knowledge" took center stage. See Luke Wilson, "William Harvey's 'Prelectiones': The Performance of the Body in the Renaissance Theater of Anatomy," *Representations* 17 (1987):62–95. See also chapter 4 of Jonathan Sawday, *The Body Emblazoned: Dissection and the Human Body in Renaissance Culture* (London: Routledge, 1995).

38 The circus performance took place in July 1999, and was filmed by VPRO television (Netherlands); part of this performance and the quoted interview with von Hagens were broadcast in the program *Het Eeuwige Lichaam* (The Eternal Body) on VPRO public television, November 28, 1999.

39 The German painter Joseph Beuys is famous not only for his still works of art, but particularly for his artistic performances, in which he often called attention to the similarities between technological and cultural tools and materials. In a performance in a New York museum, in the 1970s, he locked himself in a cage with a coyote, where he stayed for seven days. Within several days, the coyote and the artist were sleeping on a blanket together, symbolizing the fraternization of humans and wild animals.

40 *Körperwelten,* video (Heidelberg: institüt für Plastination 1997).

41 The public dissection took place on November 20, 2002, and was performed on the corpse of a seventy-two-year-old German donor, who had given permission for this act. The public "anatomy lesson" was recorded by the commercial station Channel 4, and Scotland Yard sent in two agents to determine whether this public event was in fact a medical autopsy or an artistic performance.

42 Because of the anonymity of the cadavers, their provenance is somewhat shady. Von Hagens claims that he obtains all his cadavers through private donations, yet questions have been raised about some bodies imported from Asia. Although the anatomist-artist guarantees the anonymity of each body, he admits in interviews that he himself can identify each plastinate, and that some of these cadavers were his friends or relatives while alive. A recent article in *Der Spiegel* cast serious doubts on von Hagens's "trade practices" with Asian countries. See Sven Röbel and Andreas Wassermann, "Händler des Todes," *Der Spiegel* 4 (2004): 36–50.

43 In the documentary "Die Leichenshow: Eine Ausstellung wird Sensation," directed by Walter Soechel (ARD, April 1998), various people voiced their opinions; commentators included the director of the Anatomisch-Pathologisches Bundesmuseum in Vienna and an anatomist related to the famous Fragonard Museum in Paris.

44 For instance, Marc Quinn's model of a human head filled with his own blood; the French artist Orlan's surgical performances (1991–98), in which she refashions her own body into an organic collage of artistic representations (from the forehead of Leonardo's Mona Lisa to the chin of Botticelli's *Venus*); and the severed head of an executed Maori warrior auctioned at the European Fine Art Fair (1997). The exhibition Sensation: Young British Artists from the Saatchi Collection (London, 1997; New York, 1999) featured some artists that produce body art, such as Mona Hatoum, Marc Quinn, and others. For an introduction to the discussion on the ethicality of these art works, see Norman Rosenthal, "The Blood Must Continue to Flow," in the catalogue accompanying this exhibition (London: Royal Academy of Arts, 1999), no page numbers.

45 In many respects, the discussion triggered by Bodyworlds fits the ongoing debate in contemporary philosophical and cultural criticism concerning the plasticity and manipulability of (especially female) bodies. See, for instance,

Elizabeth Grosz, *Volatile Bodies: Toward a Corporeal Feminism* (New York: Routledge, 1994), and Anne Balsamo, *Technologies of the Gendered Body: Reading Cyborg Women* (Durham, N.C.: Duke University Press, 1996).

46 N. Katherine Hayles, *How We Became Posthuman: Virtual Bodies in Cybernetics, Literature, and Informatics* (Chicago: University of Chicago Press, 1999), 5.

CHAPTER 4
FANTASTIC VOYAGES IN THE AGE OF ENDOSCOPY

1 *Fantastic Voyage,* Richard Fleischer, dir. (Columbia Pictures, 1966).

2 For a very interesting analysis of the movie from a cultural studies perspective, see Kim Sawchuk, "Biotourism, Fantastic Voyage, and Sublime Inner Space," in *Wild Science: Feminist Images of Medicine and Body,* ed. Janine Marchessault and Kim Sawchuk (New York: Routledge, 2000), 9–23.

3 Jonathan Sawday, *The Body Emblazoned: Dissection and the Human Body in Renaissance Culture* (London: Routledge, 1995), explains how in the sixteenth and seventeenth centuries, anatomists identified themselves with contemporary explorers like Columbus; they cut open bodies to discover the proper locations and functions of organs, and, like geographical cartographers, labeled the newly discovered body parts with their own names, such as Fallopius or Eustachius.

4 Laura Mulvey, in her famed article "Visual Pleasures and Narrative Cinema," *Screen* 16 (1975): 6–18, posits that cinematic spectatorship is divided along lines of gender. A male gaze either voyeuristically investigates or fetishizes a female object. Narrative and spectacle are determined by the demands of the patriarchal unconscious. The endoscopic gaze, I argue here, does not follow the vector of spectatorship proposed by Mulvey; it rather forces the viewer's gaze to an impersonal, desexualized, and therefore degendered, insider's point of view. Obviously, I concur with Mulvey about the gendered axes of spectatorship, and to underscore her point I will consistently refer to a female surgeon and male patient when talking about doctors and patients in general. Later in this chapter, I will show how the clinical gaze of the endoscope erases the (gendered) boundaries of the object.

5 Laurits Lauridsen, *Lanterna Magica in Corpore Humano* (Arhus: Steno Museum, 1998), provides an excellent introduction to the history of endoscopy.

6 Stefan Hirschauer, "The Manufacture of Bodies in Surgery," *Social Studies of Science* 21 (1991): 279–319. Quotation on page 299.

7 As Hirschauer argues in "The Manufacture of Bodies," the scene would be totally different if the patient were conscious: "A body cut open and laid bare internally—with organs hanging out or dragging out—is more than naked. Its inhabitant would be seized with fear and dismay, but would also react with a different social affect already required for states of lesser disarray of one's appearance: shame" (305). For a similar view, see Karen Dale, "Identity in a Culture of Dissection: Body, Self and Knowledge," in *Ideas of Difference: Social Spaces and the Labour of Division,* ed. Kevin Hetherington and Rolland Munro (London: Blackwell, 1997), 94–113, who regards the scalpel and the mirror as two sym-

bols for, respectively, fear and the desire to see one's own interior: "The scalpel indicates the pervasiveness of the desire to cut beneath the surface, to make visible what is hidden, to be incisive. On the other hand, the mirror indicates our desire to know ourselves, to be self-reflexive, to have a coherent idea of our own identity. Yet to see ourselves from the inside is a profoundly disturbing activity, for to observe our insides—intestines, viscera and organs—is to face the personal in ways which presage death and disorder" (99).

8 Laurits Lauridsen, in *Lanterna Magica,* 72–82, describes the precise development of these three solutions.

9 See, for instance, B. Sherman, "Techniques Tomorrow," *Modern Photography* 36, March 1972; "Dr. Berry: World-Renowned Expert on Endoscopy," *Ebony* 30, June 1975; and "Testing Fetuses" (on fetoscopy), *Time* 115, 24 March 1980.

10 B. I. Hirschowitz, "Development and Application of Endoscopy," *Gastroenterology* 104 (1993): 337–42, gives a more medical-technical description of the history of endoscopy.

11 Stated by Lauridsen, *Lanterna Magica,* 97.

12 Yutaka Yoshinaka made this point in "Exploring the Boundaries of Technological Practice In-the-Making" (paper presented at the Demarcation Socialised Conference, Cardiff, August 2000).

13 Ella Shohat, "Lasers for Ladies: Endo Discourse and the Inscription of Science," in *The Visible Woman: Imaging Technologies, Gender, and Science,* ed. Paula Treichler, Lisa Cartwright, and Constance Penley (New York: New York University Press, 1998), 240–72, makes and substantiates this claim.

14 The use of video-steered instruments may even promote interaction and discussion between doctors and specialists, argues Bonnie Nardie, "Video-as-Data: Technical and Social Aspects of a Collaborative Multimedia Approach," in *Video-Mediated Communication,* ed. Abigail Sallen, Sylvia Wilbur, and Kathleen Fink (Mahwah, N.J.: Lawrence Elbaum, 1997), 487–517.

15 On the legal consequences of endoscopic records, see P. D. Gerstenberger and P. A. Plumeri, "Malpractice Claims in Gastrointestinal Endoscopy: Analysis of an Insurance Industry Data-base," *Gastrointestinal Endoscopy* 39 (1993): 132–38; and J. Natali, "Medicolegal Implications of Vascular Injuries during Videoendoscopic Surgery," *Journal Des Maladies Vasculaires* 21 (1996): 223–26.

16 The medical talk show *Surgeon's Work (Chirurgenwerk)* started in 1990, and has been broadcast weekly by the Evangelical Broadcast Organization (Evangelische Omroep), the most conservative Dutch public broadcast organization. The talk shows are produced by René Stokvis; the particular program discussed here was aired in 1997.

17 A. Cushieri and G. Berci formulate this warning in *Laparoscopic Biliary Surgery,* (Oxford: Blackwell, 1990), ix, 109.

18 For a discussion on this issue of lowering thresholds for endoscopic surgery, see, for instance, F. Froehlich and J. J. Gonvers, "Gastrointestinal Endoscopy: Do We Perform Too Many or Not Enough Procedures?" *Canadian Journal of Gastroenterology* 13 (1999): 345–46; and L. Seematter-Bagnoud and J. Vader, "Overuse and Underuse of Diagnostic Upper Gastrointestinal Endoscopy in Various Clinical Settings," *International*

Journal for Quality in Health Care 11 (1999): 301–8.

19 See Richard M. Satava, *Cybersurgery: Advanced Technologies for Surgical Practice* (New York: Wiley-Liss, 1998); and Richard A. Robb and S. Aharon, "Patient-specific Anatomic Models from 3-dimensional Medical Image Data for Clinical Applications in Surgery and Endoscopy," *Journal of Digital Imaging* 10 (1997): 31–35.

20 Tim Lenoir, "The Virtual Surgeon," *Cahier de Science et Vie* (October 1999): 52–59, minutely describes how a surgeon's work will change in a new operative environment, especially when the ARPA system and Phantom system, two robot-operated digital systems now already experimentally used, will be fully implemented. See also E. L. Grimson, "Image-Guided Surgery," *Scientific American* (June 1999): 62–69, for a clear technical description of these high-tech environments.

21 Satava, *Cybersurgery,* 22.

22 Satava, *Cybersurgery,* 142.

23 Satava, *Cybersurgery,* 142.

24 For more technical details on the practice of telesurgery, see P. Garner and M. Collins, "The Application of Telepresence in Medicine," in *Telepresence,* ed. P. J. Sheppard and G. R. Walker (Boston: Kluwer, 1999), 323–33. See also J. Marescaux and D. Mutter, "The Virtual University Applied to Telesurgery: From Tele-education to Telemanipulation," *Bulletin de L'Académie Nationale de Médecine* 183 (1999): 509–22.

25 Satava, *Cybersurgery,* 11.

26 So far, comparisons between virtual and video endoscopy have been limited to the domain of colonoscopy. See, for instance, Helen M. Fenlon, "A Comparison between Virtual and Conventional Colonoscopy for the Detection of Colorectal Polyps," *New England Journal of Medicine* 341 (November 11, 1999): 1496–503; see also Martina M. Morrin, "Virtual Colonoscopy: A Kinder, Gentler Colorectal Cancer Screening Test?" *Lancet* 354 (September 25, 1999): 1048–49. On the usefulness of virtual endoscopy for surgical education and training, see P. J. Gorman and A. H. Meier, "Simulation and Virtual Reality in Surgical Education—Real or Unreal?" *Archives of Surgery* 134 (November 1999): 1203–8.

27 At this stage in time, virtual endoscopy is far from ready for general screening purposes, according to most researchers, such as John H. Bond, "Virtual Colonoscopy—Promising but Not Ready for Widespread Use," *New England Journal of Medicine* 341: (November 11, 1999): 1540–42.

28 William J. Mitchell, *The Reconfigured Eye: Visual Truth in the Photographic Era.* (Cambridge, Mass.: MIT Press, 1992). See particularly 163–66 on "mise-en-image." A more radical stance is taken by Eugene Thacker, "What Is Biomedia?" *Configurations* 11 (2003): 47–79, who argues that informatics and biology (and thus their respective representations) are intricately intertwined in contemporary biomedia, especially in relation to bioengineering and genetics: "Biomedia is neither a technological instrument, nor an essence of technology, but a situation in which a technical, informatic recontextualization of biological components and processes enables the body to demonstrate itself in demonstrations that may be biological or non-biological, medical or militaristic, cultural or economic" (79).

29 *Body Story.* An eight-part series pro-

duced by BBC Television, 1998. Broadcast by various European broadcast organizations in 1998 and 1999.

30 Steve Connor, "Integuments: The Scar, the Sheen, the Screen," *New Formations* 39 (1999): 32–54. Citation on page 52.

31 Body Wars, an attraction at Epcot at Walt Disney World, Florida, as described on its Web site: "Shrink down to the size of a blood cell and take a turbulent ride through the human body. This flight-simulator-style attraction races through a larger-than-life version of the heart, lungs, and brain on a mission to rescue a scientist trapped in inner space."

32 Erkki Huhtamo, "Encapsulated Bodies in Motion," in *Critical Issues in Electronic Media,* ed. Simon Perry (New York: SUNY Press, 1995), 159–86. Citation on page 164.

33 Jay D. Bolter and Richard Grusin, *Remediation: Understanding New Media* (Cambridge, Mass.: MIT Press, 1998), 166.

34 For an interesting analysis of digital instruments (particularly 3-D ultrasound) and the philosophical-ideological framework for addressing questions of conceptualizing the body, see Karen Barad, "Getting Real: Technoscientific Practices and the Materialization of Reality," *Differences* 10 (1998): 87–128.

35 Tim Lenoir and Xin Wei Sha, "Authorship and Surgery: The Shifting Ontology of the Virtual Surgeon," in *From Energy to Information,* ed. Bruce Clark and Linda Henderson (Stanford: Stanford University Press, 2002), suggest that the implementation of new virtual technologies will allow people other than the surgeon to have power over the surgical body.

36 Erkki Huhtamo, in "Encapsulated Bodies," argues that "technology is gradually becoming a second nature, a territory both external and internalized, and an object of desire" (171).

37 Anne Balsamo, *Technologies of the Gendered Body: Reading Cyborg Women* (Durham: Duke University Press, 1996), points out that polished images may appear to render bodies transparent, yet this does not automatically translate into better diagnoses or treatment. More than anything, vivid images create higher expectations and norms vis-à-vis the manipulable body: "The fact that virtual realities offer new information environments does not guarantee that people will use the information in better ways" (132).

38 On the influence of viewers' aesthetics on patients' standards, see Eugene Thacker, "Performing the Technoscientific Body: Real Video Surgery and the Anatomy Theater," *Body and Society* 5 (1999): 317–36.

39 Pierre Lévy, *Becoming Virtual: Reality in the Digital Age* (New York: Plenum Trade, 1998), suggests that scanners and other medical imaging devices erase the human skin but concurrently create new layers: "Each new device adds another type of skin, another visible body to our actual body. The organism is turned inside out like a glove. The interior appears on the outside, while remaining within. For the skin is also the boundary between the self and the external world" (40).

40 Henry Fountain, "'Camera in a Pill' Views Digestive Tract," *New York Times,* May 30, 2000.

CHAPTER 5
X-RAY VISION IN THOMAS MANN'S *THE MAGIC MOUNTAIN*

1 Marcel Proust, *Swann's Way,* vol. 1 of *Remembrance of Things Past,* trans. C. K.

Scott Moncrieff (London: Chatto and Windus, 1943), 70. Originally published as *Du côté de chez Swann* in 1924 as part of *A la Récherche du Temps Perdu.*

2 In the 1870s, the French doctor Claude Bernard propagated the "experimental method," a systematic approach to scientific research for the disciplines of medicine and biology based on Auguste Comte's and Hyppolyte Taine's positivism, which called for an explanation of phenomena solely on the basis of facts, excluding speculation and promoting rationalization. Bernard's ideas, in turn, formed the basis for Emile Zola's "Le Roman Experimental" (1880), in which Zola applied the experimental method to literature. Zola's essay became the manifesto for the naturalist movement.

3 It should be noted that the term "objectivity" has a specific historical meaning. Lorraine Daston, "Objectivity and the Escape from Perspective," *Social Studies of Science* 22 (1992): 597–618, distinguishes three meanings of nineteenth-century objectivity: ontological, mechanical, and aperspectival objectivity. "Whereas ontological objectivity is about the world, and mechanical objectivity is about suppressing the universal human propensity to judge and to aestheticize, aperspectival objectivity is about eliminating individual . . . idiosyncrasies" (599). Although I will not elaborate on the historical variability of the definition of objectivity, for the remainder of this chapter I will assume the term refers to the last option in Daston's classification: aperspectival objectivity.

4 For a general description of the way new instruments changed the doctor's relation to the senses in nineteenth-century diagnostics, see Stanley Reiser, "Technology and the Use of the Senses in Twentieth-century Medicine," in *Medicine and the Five Senses,* ed. Catherine F. Byrum and Roy Porter (Cambridge: Cambridge University Press, 1996), 262–73.

5 Thomas L. Hankins and Robert J. Silverman, *Instruments and the Imagination* (Princeton: Princeton University Press, 1995), describe a number of interrelated examples of visual, graphic, and acoustic recording technologies in the second half of the nineteenth century. They observe that it was unclear, at the moment of their inventions, whether these technologies belonged to a scientific armamentarium or to the domain of arts and entertainment.

6 Lisa Cartwright, *Screening the Body: Tracing Medicine's Visual Culture* (Minneapolis: University of Minnesota Press, 1995), 107.

7 Thomas Mann, *The Magic Mountain,* trans. Helen Tracy Lowe-Porter (New York: Random House, 1969); originally published as *Der Zauberberg,* (Fischer Verlag, 1924).

8 For an extensive historical description of how the epidemic of tuberculosis in Great Britain led to a medical and social infrastructure, see Linda Bryder, *Below the Magic Mountain: A Social History of Tuberculosis in Twentieth-Century Britain* (Oxford: Clarendon Press, 1988).

9 Barron H. Lerner, "The Perils of X-Ray Vision: How Radiographic Images Have Historically Influenced Perception," *Perspectives in Biology and Medicine* 35 (1992): 382–97, provides an interesting analysis of the period during which medical diagnostic interpretations of X rays were still in flux.

10 Bryder, *Below the Magic Mountain,* 105.

11 Mann, *Magic Mountain,* 178.

12 Mann, *Magic Mountain,* 186.

13 Barbara Bates, *Bargaining for Life: A Social History of Tuberculosis, 1876–1938* (Philadelphia: University of Pennsylvania Press, 1992), provides a thorough history of social life in sanatoriums in the early part of the twentieth century.

14 For an excellent introduction to the social construction of tuberculosis as a disease and the X ray as a diagnostic instrument, see Bernike Pasveer, "Knowledge of Shadows: The Introduction of X-Ray Images in Medicine," *Sociology of Health and Illness* 11 (1989): 360–81.

15 Bernike Pasveer, in "Knowledge of Shadows," extensively explains the process of consensus-building using comparisons between "older" and "newer" diagnostic methods. See also T. Hugh Crawford, "Imaging the Human Body: Quasi Objects, Quasi Texts, and the Theater of Proof," *PMLA* 111 (1996): 66–79.

16 Lerner, "Perils of X-ray Vision," 393.

17 Mann, *Magic Mountain,* 242.

18 Mann, *Magic Mountain,* 625.

19 Thomas Hankins and Robert Silverman, *Instruments and the Imagination,* elaborate on the kinship between the gramophone and other recording instruments, including graphical and audio inscription devices, such as the pentagraph and the *vox mechanica.* Whereas Hankins and Silverman's book is more or less a historiography of instruments, Friedrich A. Kittler's *Gramophone, Film, Typewriter,* trans. Geoffrey Winthrop-Young and Michael Wutz (Stanford: Stanford University Press, 1999), (originally published in 1986 as *Grammophon Film Typewriter*), also encompasses the social uses and validations of various representational instruments (audio, image, text), demonstrating how they gradually changed the cultural landscape in the nineteenth century. According to Kittler, the gramophone reproduced the "original voice"; for the first time, a piece of music could be listened to multiple times in exactly the same performance. Before the emergence of the gramophone, every performance was unique and only the notation system was standardized.

20 Mann, *Magic Mountain,* 642.

21 Of course, it would be another ten years after the publication of *The Magic Mountain* before Walter Benjamin would publish his now-famous essay "Art in the Age of Mechanical Reproduction" (1935); Thomas Mann's ideas on the reproduction of original art clearly anticipate Benjamin's.

22 Friedrich Kittler, in *Gramophone, Film, Typewriter,* emphasizes this relationship between the gramophone and other representational instruments much more explicitly. He describes how, in 1927, Thomas Mann, upon being asked whether *The Magic Mountain* could be turned into a movie, replied that the novel in fact was already close to a scenario. Images, dialogues, and music were described in such detail that they were "ready to be recorded"—the various representations only had to be combined in the medium of film (173–74).

23 For a description of popular uses of the X ray, see Bettyann Holtzmann-Kevles, *Naked to the Bone: Medical Imaging in the Twentieth Century* (New Brunswick, N.J.: Rutgers University Press, 1997), chapters 1 and 2.

24 The fluoroscope was not the only frivolous application of the X rays that was particularly popular among women. Rebecca Herzig, "Removing Roots: North American Hiroshima Maidens and the X-ray," *Technology and Culture* 40 (1999): 723–45, relates how women

visited special clinics to have abundant hair growth removed by X rays in order to obtain the smoothest skin possible. The terrible implications of this type of cosmetic treatment did not become obvious until 1918, but it wasn't until at least the 1940s that all clinics that used this treatment were closed down.

25 Lisa Cartwright, *Screening the Body,* 102–24, provides an extensive description of X ray's ascribed psychological and erotic power.

26 Cartwright, *Screening the Body,* 115.

27 Mann, *Magic Mountain,* 216.

28 Susan Sontag, *Illness as Metaphor* (New York: Farrar, Straus and Giroux, 1977).

29 For more interesting details on the romances between patients in sanatoriums, see Sheila M. Rothman, *Living in the Shadow of Death: Tuberculosis and the Social Experience of Illness in American History* (New York: Harper, 1994).

30 Mann, *Magic Mountain,* 342.

31 Mann, *Magic Mountain,* 341.

32 Mann, *Magic Mountain,* 348.

33 Mann, *Magic Mountain,* 610.

34 Stanley J. Reiser, *Medicine and the Reign of Technology* (Cambridge: Cambridge University Press, 1978), 58–68.

35 Holtzmann-Kevles, *Naked to the Bone,* 59.

36 Mann, *Magic Mountain,* 257.

37 Mann, *Magic Mountain,* 259.

38 Mann, *Magic Mountain,* 260.

39 Mann, *Magic Mountain,* 218–19.

40 For a wonderful introduction to the emergence of the Society for Psychical Research, see Alan Grove, "Röntgen's Ghosts: Photography, X-Rays, and the Victorian Imagination," *Literature and Medicine* 17 (1998): 141–73.

41 Grove, "Röntgen's Ghosts," 164.

42 This argument is made, for instance, by T. J. Reed in *Thomas Mann: The Uses of Tradition* (Oxford: Oxford University Press, 1974).

43 For this type of explanation, see John S. King, "Most Dubious: Myth, the Occult, and Politics in the Zauberberg," *Monatshefte* 88 (1996): 217–36.

44 Mann, *Magic Mountain,* 659.

45 Mann, *Magic Mountain,* 666.

46 Mann, *Magic Mountain,* 667.

47 Mann, *Magic Mountain,* 673.

48 Mann, *Magic Mountain,* 680. Friedrich Kittler, in *Gramophone, Film, Typewriter,* observes that Ziemssen's ghost does not appear until they start playing the gramophone in the séance room: "The sanatorium's own psychoanalyst is unable to conjure up the spirit of Castorp's deceased cousin until the gramophone administrator comes up with the obvious solution. Only when prompted by the phonographic reproduction of his favorite tune does the spirit appear, thus revealing this media link to be a sound-film reproduction" (174).

49 For Thomas Mann's thoughts on parapsychology and occultism, see his "Drei Berichte über Okkultische Sitzungen," *Gesammelte Werke* vol. 13 (Berlin: Fisher Verlag, 1965), 33–48.

50 This argument is made by Richard Koc, "Magical Enactments: Reflections on 'Highly Questionable' Matters in *Der Zauberberg,*" *Germanic Review* 68 (1993): 107–17.

51 Mann, *Magic Mountain,* p. 84.

52 Jonathan Crary, *Techniques of the Observer: on Vision and Modernity in the Nineteenth Century* (Cambridge, Mass.: MIT Press, 1990), argues that the emergence of a new visual regime is directly related, yet not caused by, a complex of technological devices.

CHAPTER 6
ULTRASOUND AND THE VISIBLE FETUS

1 This chapter is based on a review of Dutch national regulations on ultrasound, regulatory protocols, and insurance policies in the past fifteen years. In addition, I conducted twenty-two in-depth interviews with Dutch gynecologists, midwives, general practitioners, and pregnant women between June 1999 and January 2000. My research took place in one particular region of the Netherlands (the area of Maastricht), but the situation in this area is indicative of regulatory changes nationwide. A similar debate is taking place in Britain; see, for instance, Trish Chudleigh, "Scanning for Pleasure," *Ultrasound in Obstetrics and Gynaecology* 14 (1999): 369–71; and C. Baillie, G. Mason, and J. Hewison, "Scanning for Pleasure," *British Journal of Obstetrics and Gynaecology* 104 (1997): 1223–24. However, a meaningful comparison would require a careful scrutiny (and taking into account) of the different health systems, which is beyond the scope of this chapter. I would like to thank all interviewed professionals and parents, who substantially increased my understanding of the (changing) practices of prenatal ultrasound in the Netherlands.

2 For an interesting reading of the complex conjunctions between bodies, machines, and trajectories, see Bernike Pasveer, "Wiens lijf? Wiens Leven? Echografie en het lichaam," *De Nieuwe Mens: De Maakbaarheid van Lijf en Leven,* ed. F. de Lange (Nijmegen: Thomas Moore Academie, 2000), 43–57; see also Marc Berg and Annemarie Mol, eds., *Differences in Medicine: Unraveling Practices, Techniques, and Bodies* (Durham: Duke University Press, 1998).

3 For a description of the theory of social construction of technology, see Wiebe Bijker and John Law, *Shaping Technology/Building Society* (Cambridge, Mass.: MIT Press, 1992), particularly pages 17–19.

4 Paul G. Newman and Grace S. Rozycki, "The History of Ultrasound," *Surgical Clinics of North America* 78 (1998): 179–95, argue that the history of ultrasound does not start in the twentieth but rather in the nineteenth century. In 1826, the Swiss physician Jean-Daniel Colladon started experimenting with SONAR (SOund Navigation And Ranging); Christian Andreas Doppler, an Austrian physicist and mathematician, articulated his "Doppler theory" on the various frequencies of sound as early as 1841.

5 For an interesting description of the history of ultrasound, see Edward Yoxen, "Seeing with Sound: A Study of the Development of Medical Images," *The Social Construction of Technological Systems: New Directions in the Sociology and History of Technology,* ed. Wiebe Bijker, Thomas Hughes, and Trevor Pinch (Cambridge, Mass.: MIT Press, 1987), 281–303.

6 An extensive and detailed history of the technical-medical use of ultrasound is provided by Margareth B. McNay and John E. Fleming, "Forty Years of Obstetric Ultrasound, 1957–97: From A-scope to Three Dimensions," *Ultrasound in Medicine and Biology* 25 (1999): 3–56, who observe that "visualization of the fetus had been found more or less by accident when examining a case thought clinically to have uterine enlargement due to fibromyomata" (8).

7 Stuart Blume, *Insight and Industry: On the Dynamics of Technological Change in Medicine* (Cambridge, Mass.: MIT Press, 1992), describes the socio-technological history of various inter-related imaging technologies.

8 The professional struggle between gynecologists, radiologists, and surgeons is carefully charted by Ellen Koch, "In the Image of Science? Negotiating the Development of Diagnostic Ultrasound in the Cultures of Surgery and Radiology," *Technology and Culture* 34 (1993): 858–93.

9 Pregnant women who have to subject themselves to a transabdominal ultrasound are usually asked to drink two glasses of water before the exam; in the case of a vaginal ultrasound exam, where the transducer is inserted through the vagina, a full bladder is not necessary to obtain sharp images.

10 McNay and Fleming, "Forty Years of Obstetric Ultrasound," 26.

11 See Holtzmann-Kevles, *Naked to the Bone,* 243–50.

12 Prices of ultrasound equipment vary greatly: the cheapest machines cost as little as $10,000 to $15,000; ultrasound computers, which sonographers in clinics use to detect subtle aberrations, such as heart and kidney failures, may cost as much as $300,000. The newest three-dimensional ultrasound equipment is even more expensive.

13 I did not find any statistical evidence for the number of fetal defects detected by means of ultrasound; the gynecologists I interviewed estimated the detection potential of ultrasound at 60 or 70 percent at the maximum.

14 K. Ammann and K. Knorr Cetina, "The Fixation of (Visual) Evidence," in *Representation in Scientific Practice.* ed. Michael Lynch and Steve Woolgar (Cambridge, Mass.: MIT Press, 1988), 85–122. Citation on page 90.

15 McNay and Fleming, "Forty Years of Obstetric Ultrasound," pay attention to the tandem development of ultrasound techniques and interpretive skills; they conclude that the developmental stage of an imaging technology is always confusing for both doctors and patients, and consensus only emerges gradually after years of contested interpretation. "Questions arose such as 'what is the significance of the presence of the isolated choroid plexus cyst at 18 weeks or nuchal thickening in the first trimester?'" (44).

16 In the case of X rays, as described in the previous chapter, it took several decades. This process is described minutely by Bernike Pasveer, "Knowledge of Shadows: The Introduction of X-ray Images in Medicine," *Sociology of Health and Illness* 11 (1989): 360–81.

17 Lorna Weir, "Cultural Intertexts and Scientific Rationality: The Case of Pregnancy Ultrasound," *Economy and Society* 27 (1998): 249–58. Citation on page 250.

18 Listening to the fetus is still an important part of the prenatal checkup, usually with the help of Doppler equipment, which enables a constant monitoring of the fetal heartbeat. For a description of the use of medical equipment during pregnancy and delivery, see Madeleine Akrich and Bernike Pasveer, "Passages of Surveillance and Coordination," *Theoretical Medicine* 21 (2000), 63–83.

19 Anne Balsamo, *Technologies of the Gendered Body: Reading Cyborg Women* (Durham: Duke University Press, 1996), chapter 4, has commented extensively upon the cultural impact of ultrasound.

20 Both Rosalind Petchesky, *Abortion and Woman's Choice: The State, Sexuality, and Reproductive Freedom* (Boston: North-

eastern University Press, 1984), and Irma van der Ploeg, *Prosthetic Bodies. Female Embodiment in Reproductive Technologies* (Maastricht: Maastricht University Press, 1998), have commented on the autonomous status of the fetus in conjunction with ultrasound practices.

21 See Beverly Hyde, "An Interview Study of Pregnant Women's Attitudes to Ultrasound Screening," *Social Science and Medicine* 22 (1986): 587–92.

22 This claim is backed up by S. G. Gabbe, "Routine versus Indicated Scans," in *Diagnostic Ultrasound Applied to Obstetrics and Gynaecology,* ed. Ruddy E. Sabbagha (Philadelphia: Lippincott, 1994).

23 Yona Teichmann and Dorit Rabinovitz, "Emotional Reactions of Pregnant Women in Ultrasound Scanning and Postpartum," in *Stress and Anxiety,* ed. Charles Spielberger and Irwin G. Sarason (New York: Hemisphere Press, 1991), claim that clinical information on the health status of the fetus reduces emotional stress; on the other hand, S. Ayers and A. D. Pickering, "Psychological Factors and Ultrasound: Differences between Routine and High-risk Scans," *Ultrasound in Obstetrics and Gynaecology* 9 (1997): 76–79, argue that ultrasound scans tend to increase psychological pressure.

24 See S. Campbell and E. Reading, "Ultrasound Scanning in Pregnancy: The Short-term Psychological Effects of Early Real-time Scans," *Journal of Psychosomatic Obstetrics and Gynaecology* 1 (1982): 57–60.

25 J. C. Fletcher and M. I. Evans, "Maternal Bonding in Early Fetal Ultrasound Examinations," *New England Journal of Medicine* 308 (1983): 392–93, observe a causal relationship between ultrasound and prenatal bonding. This research was rooted in the claim made eleven years earlier by M. Klaus and P. Jerauld, "Maternal Attractions: Importance of the First Postpartum Days," *New England Journal of Medicine* 286 (1972): 460–63, that the intimate contact between mother and child immediately after birth has very positive effects on bonding. However, as Janelle C. Taylor correctly observes in "Image of Contradiction: Obstetrical Ultrasound in American Culture," in *Reproducing Reproduction, Kinship, Power and Technological Innovation,* ed. Sarah Franklin and Helena Ragone (Philadelphia: University of Pennsylvania Press, 1997), 15–45, the maternal affection of pregnant women for their fetuses, as described by Klaus, cannot simply be equated with women's affection for their newborns; the emotional bond between mother and baby is radically different from the same bond in the prenatal condition.

26 C. Baillie, "Should Utrasound Scanning in Pregnancy Be Routine?" *Journal of Reproductive and Infant Psychology* 17 (1999): 149–57.

27 J. A. M. Hunfeld, "Emotional Reactions in Late Pregnancy Following the Ultrasound Diagnosis of Severe or Lethal Fetal Malformation," *Prenatal Diagnosis* 13 (1993): 603–12.

28 Martin P. Johnson and John E. Pudifoot, "Miscarriage: Is Vividness of Visual Imagery a Factor in the Grief Reaction of the Partner?" *British Journal of Health Psychology* 3 (1998): 137–46.

29 Taylor, "Image of Contradiction," 24.

30 It is beyond the scope of this book to provide an exhaustive list of references that balance the positive and negative effects of ultrasound. To name just a few often-quoted examples: Rita B. Black, "Seeing the Baby: The Impact of Ultra-

sound Technology," *Journal of Genetic Counseling* 1 (1992): 45–54; Thomas R. Verny, "Obstetrical Procedures: A Critical Examination of Their Effect on Pregnant Women and Their Unborn and Newborn Children," *Pre- and Peri-Natal Psychology Journal* 7 (1992): 101–12; and Valerie D. Raskin, "Influence of Ultrasound on Parent's Reaction to Perinatal Loss," *American Journal of Psychiatry* 146 (1989): 1646.

31 This claim is substantiated by Josephine Green and Helen Statham, "Psychosocial Aspects of Prenatal Screening and Diagnosis," in *The Troubled Helix: Social and Psychological Implications of the New Human Genetics,* ed. Theresa Marteau and Martin Richards (Cambridge: Cambridge University Press, 1996), 140–58.

32 This claim is made by Beverly Hyde in "Interview Study of Pregnant Women," 591.

33 There is a distinct asymmetry concerning the implementation of prenatal ultrasound in the various countries surrounding the Netherlands. For instance, the ultrasound prenatal scan is a routine screening in Germany and France, where one scan per trimester is the norm; Belgium and Great Britain allow at least two scans covered by national health insurance. For an overview of various policies in European countries, see Salvatore Levi, "Routine Ultrasound Screening of Congenital Anomalies: An Overview of the European Experience," in *Ultrasound Screening for Fetal Anomalies: Is It Worth It?,* ed. Salvatore Levi and Frank Chervenak (New York: New York Academy of Sciences, 1998), 86–97.

34 The Dutch National Health Association issues standards that explicitly reject ultrasound as a screening instrument based on empirical evidence and large-scale trials. See *Nederlands Genootschap voor Huisartsen, Standaarden voor de Huisarts II* (1999), 314.

35 The claim that routine screening has done nothing to lower fetal morbidity rates is substantiated by Bernard G. Ewigman, "Effect of Prenatal Ultrasound Screening on Perinatal Outcome," *New England Journal of Medicine* 329 (1993): 821–829.

36 Some of these side effects are claimed by Helle Kieler, *Effects and Possible Side Effects of Routine Ultrasound Scanning in Pregnancy* (Uppsala: Acta Universitatis Upsaliensis, 1997).

37 Interviewed midwives and gynecologists estimated the percentage of Dutch women who have at least one ultrasound during their pregnancy to be between 95 and 98 percent.

38 For an interesting comparison between the French and Dutch trajectories of pregnancy and delivery, see Akrich and Pasveer, "Obstetrical Trajectories."

39 Margarete Sandelowski, "Separate, but Less Unequal: Fetal Ultrasonography and the Transformation of Expectant Mother/Fatherhood," *Gender and Society* 8 (1994): 230–45, has described the ritualized aspects of the first sonogram; Marveen Craig, "Controversies in Obstetric Gynecologic Ultrasound," in *Diagnostic Medical Sonography: A Guide to Clinical Practice,* ed. M. Berman (Philadelphia: Lippincott, 1991), also points out how the first ultrasound is turned into a familial event.

40 High-risk pregnancies are the pregnancies of women who have a family history of genetic or congenital defects, multiparous women with previous complicated pregnancies or deliveries, and primiparous women over thirty-six.

41 Not surprisingly, some Dutch gynecologists have joined the European lobby to implement second-trimester ultrasound as routine screening, even though its effectiveness is far from proven and the debate is still ongoing. Proponents and opponents of routine second-trimester ultrasound have come up with research results to back up their views. For a proponent's view, see, for instance, D. W. Skupsi, F. A. Chervenak, and M. McCullough, "Routine Obstetric Ultrasound," *International Journal of Gynaecology and Obstetrics* 50 (1995): 233–42. For the opposing argument, see Bernard G. Ewigman and J. P. Crane, "A Randomized Trial of Prenatal Ultrasonographic Screening: Impact on Maternal Management and Outcome," *American Obstetrics and Gynaecology* 169 (1993): 483–89; and R. W. Berkowitz, "Should Every Pregnant Woman Undergo Ultrasonography?" *New England Journal of Medicine* 329 (1993): 874–75. See also A. Saari-Kemprainen and O. Karjalainen, "Ultrasound Screening and Perinatal Mortality: Controlled Trial of Systematic One-Stage Screening in Pregnancy: The Helsinki Ultrasound Trial," *Lancet* 336 (1990): 387–91; and Patril Romano and Norman J. Waitzman, "Can Decision Analysis Help Us Decide Whether Ultrasound Screening for Fetal Abnormalities is Worth It?" in Levi and Chervenak *Ultrasound Screening for Fetal Anomalies,* 154–68, for opposing arguments in this discussion.

42 In an interview, "fun-sonographer" Jeanne Offermans, who runs a private "studio" named "Echo Extra" in Brunssum, the Netherlands, explicitly underscored her medical status, stating that her clients do not just pay to have a video made, but for a "complete ultrasound checkup." Interview in *Baby's Digest* (infomercial leaflet on baby food, Friesland Nutrition, July 1999), 2–3.

43 The newest generation of high-resolution scanners allows detection of chromosomal abnormalities and structural anomalies, raising both new ethical issues and psychological dilemmas. For a discussion, see the editorial by Anne McFayden, "First Trimester Ultrasound Screening Carries Ethical and Psychological Implications," *British Medical Journal* 317 (198): 694–95.

44 This argument is made by Lorna Weir in "Cultural Intertexts and Scientific Rationality," pp. 256–58.

45 Taylor, "Image of Contradiction," 26.

46 See "The Womb as Photo Studio," technology section, *New York Times,* 23 September 2004, digital edition. The new companies have names such as Baby Insight, Peek-a-Boo Ultrasound, Womb with a View, Prenatal Peek, and Fetal Fotos.

47 Frank A. Chervenak and Laurence B. McCullough, "Ethical Dimensions of Ultrasound Screening for Fetal Anomalies," in Levi and Chervenak, *Ultrasound Screening for Fetal Anomalies,* 185–90, argue for a protocol in which medical information is strictly separated from either the doctor's or the patient's subjective values.

48 Francis Price, "Now You See It, Now You Don't: Mediating Science and Managing Uncertainty in Reproductive Medicine," in *Misunderstanding Science? The Public Reconstruction of Science and Technology,* ed. Alan Irwin and Brian Wynne (Cambridge: Cambridge University Press, 1996), 84–106, argues that medical information is never "purely" medical, but always bound up with experiential perceptions and relational and contextual knowledge.

49 Charis M. Cussins, "Ontological Choreography: Agency for Women Patients in an Infertility Clinic," in Berg and Mol, *Differences in Medicine,* 166–201. Citation on page 168.

50 For an interesting perspective on the concept of autonomy, see C. MacKenzie and N. Stoljar, *Relational Autonomy* (Oxford: Oxford University Press, 2000); they define relational autonomy as "a characteristic of agents who are emotional, embodied, desiring, creative, feeling, as well as rational creatures" who can only understand and be understood in "the rich and complex social and historical contexts in which [they] are embedded" (21).

CHAPTER 7
DIGITAL CADAVERS AND VIRTUAL DISSECTION

1 For a detailed explanation of the transformation of an anatomical body into a "body of knowledge," see Jonathan Sawday, *The Body Emblazoned: Dissection and the Human Body in Renaissance Culture* (London: Routledge, 1995), chapter 1.

2 Dissections, as part of medical education, took place as early as the fourteenth century, but it wasn't until the sixteenth century that medical practice was actually based on empirical anatomical findings—knowledge that was subsequently recorded in medical atlases. See Roger French, *Dissection and Vivisection in the European Renaissance* (Aldershot: Ashgate, 1999). The use of anatomical illustration in the Renaissance is a rather complicated issue. The aim of anatomical illustrations was never (not even in Vesalius's work) that of substituting the text or the practice of dissection, but more simply to provide an *aide de mémoire* in support of the text. The point was not that they had to be accurate, naturalistic representations, but that they had to help the students remember what they had read in the texts, or later, seen through direct observation. Over the fifteenth and sixteenth centuries there was an epistemological shift from bookish learning towards direct observation. The visual expression of this methodological and epistemological shift is evident in the representations of anatomies in the frontispiece of Ketham's *Fasciculus medicinae* and the very different frontispiece of Vesalius's *Fabrica.*

3 This claim is made by Michael J. Ackerman, "The Visible Human Project: A Resource for Education," *Academic Medicine* 74 (1999): 667–70.

4 The Center for Human Simulation's homepage provides a short overview of how the Visible Human Project evolved: http://www.nlm.nih.gov/research/visible/visible_human.html

5 The Visible Human data sets are now available on CD-ROM: *The Complete Visible Human: The Complete High-Resolution Male and Female Anatomical Databases from the Visible Human Project* (New York: Springer Verlag, 1999). It encompasses more than 7,000 cross sections and is produced for educational use.

6 As Michael J. Ackerman claims in "The Visible Human Project": "It is hoped that this Website will serve as the prototype for revolutionary educational applications based on both the core VHP data sets and additional human imagery sources to be added late[r]. . . . As the rapid proliferation of Web browsers and networking hardware has demonstrated, innovation and entrepreneurial spirit can sweep away decades of

conventional capability in a matter of months" (670).

7 Ackerman, "Visible Human Project," 668.

8 Katherine Park, "The Criminal and the Saintly Body: Autopsy and Dissection in Renaissance Italy," *Renaissance Quarterly* 1 (1994): 1–33.

9 In 1308, the body of Francesca of Foligna was cut open to see if her heart was shaped as a cross, or if a crown of thorns was hidden away in her intestines.

10 Park, "Criminal and Saintly Body," 14.

11 For an extensive description of late-medieval practices of dissection, see Andrea Carlino, *Books of the Body: Anatomical Ritual and Renaissance Learning* (Chicago: Chicago University Press, 1999), particularly chapter 1.

12 The tradition of public dissection was not a sixteenth-century invention, and thus Vesalius was by no means the first to dissect cadavers in public; public dissections were already performed in the late fifteenth century, as the city statutes of Milan, Padua, and Bologna all included regulations for the practice of anatomy as early as the late fourteenth century. Vesalius, however, was one of the most famous anatomists. The "empirical turn," naturally, did not happen instantaneously after Vesalius took center stage; initially, he used his observations to correct Galen's insights, yet incorporated these in the dominant paradigm. The empirical turn in anatomy happened gradually, but Vesalius counts as a turning point.

13 For a detailed analysis of Vesalius's techniques, see Glenn Harcourt, "Andreas Vesalius and the Anatomy of Antique Sculpture," *Representations* 17 (1987): 28–61. The issue of Vesalius's radical claims, however, has been under scrutiny in recent scholarship. See, for a more contextualized explanation, Nancy Siraisi, "Anatomizing the Past: Physicians and History in Renaissance Culture," *Renaissance Quarterly* 53 (2000): 1–30. See also Nancy Siraisi, "Vesalius and the Reading of Galen's *Teleology*," *Renaissance Quarterly* 50 (1997): 1–37.

14 Carlino, *Books of the Body*, 206–7.

15 William Harvey was famous in seventeenth-century Europe for his remarkable personality and the way in which he paired rhetorical fluency with refined dissecting techniques. See Luke Wilson, "William Harvey's 'Prelectiones': The Performance of the Body in the Renaissance Theater of Anatomy," *Representations* 17 (1987): 62–95.

16 Sawday, *Body Emblazoned*, 64.

17 Ackerman, "Visible Human Project," 668.

18 Richard Satava, for instance, claims that operations of the future will be mostly computer-directed operations in so-called virtual bodies. See "Medical Virtual Reality: The Current State of the Future," in *Health Care in the Information Age*, ed. S. J. Weghorst, H. B. Sieburg, and K. S. Morgan (Amsterdam: IOS Press, 1996): 100–106. Eugene Thacker, "Visible Human: Digital Aantomy and the Hypertexted Body," *C-Theory* 21 (1998): no page numbers, argues that the primary concern behind the VHP is as much about informatics as it is about anatomy.

19 William J. Mitchell, *The Reconfigured Eye: Visual Truth in the Photographic Era* (Cambridge, Mass.: MIT Press, 1992), argues that the eye is constantly reconditioned and reconfigured, as new digital techniques affect social and medical practices as well as institutional structures.

20 Anatomical atlases and the relationship between the artist and anatomist were very much a sixteenth-century, post-Vesalian phenomenon. Prior to Vesalius there was little systematic attempt to collect and publish anatomical illustrations. In the fifteenth century, anatomists were not concerned with illustrations that would convey their empirical findings. A large number of anatomical books, in fact, were not even illustrated. It remains questionable to what extent drawings represented empirical findings even in Vesalius and in post-Vesalian anatomy. An example of a famous Dutch collaboration between anatomist and illustrator was that of the anatomist Albinus and illustrator Jan Wandelaar. Their differences of opinion about scientific interpretations vis-à-vis representational accuracy are well documented. See the catalogue *De volmaakte mens: De anatomische atlas van Albinus en Wandelaar* (Leiden: Museum Boerhaave, 1991).

21 Catherine Waldby, in her philosophical interpretation of the VHP, "The Visible Human Project: Data into Flesh, Flesh into Data," in *Wild Science,* ed. Janine Marchessault and Kim Sawchuk (New York: Routledge, 2000) 24–38, regards the Visible Human "not so much [as] the representation of a body in space, but as a representation of bodily space, rendered as depth and volume which can be moved through and refigured at will" (33).

22 Ackerman, "Visible Human Project," 669.

23 The software made on the basis of Visible Human data sets complements rather than replaces the dissection of real cadavers, argues Paul M. Rowe, "Visible Human Project Pays Back Investment," *Lancet* 352 (1999): 46.

24 According to most users, the data sets complement conventional dissection of human cadavers. See, for instance, Katherine Rowe, "God's Handy Worke," in *The Body in Parts: Fantasies of Corporeality in Early Modern Europe,* ed. David Hillman and Carla Mazzio (London: Routledge, 1997), 285–312. Despite the claim made by the VHP, it is very difficult to learn something from digital scans without any prior knowledge. Then (as well as now) the staple of anatomical practice remained solidly anchored in words and books (as opposed to images). Even in the case of Vesalius, the printing of images was primarily a commercial venture. For a more substantial argument, see Vivian Nutton, "Representation and Memory in Renaissance Anatomical Illustration," in *Immagini per conoscere: dal Rinascimento alla Rivoluzione scientifica,* ed. Fabrizio Meroi and Claudio Pogliano (Florence: L. S. Olschki Editore, 2001), pp. 61–80.

25 A. M. Lassek, *Human Dissection: Its Drama and Struggle* (Springfield, Ill.: Thomas, 1958). It was specified both for Padua and Bologna that the corpse for the public anatomy had to be that of a criminal condemned to death. In the case of Milan the men or women simply had to be of *"vilis et humilis condicionis."* What is common to various provisions, therefore, is that the bodies had to be those of people at the fringes of society; not only criminals, but also human beings alien to the society that was to participate in the spectacle of public anatomies. Paupers and beggars that died in the Milanese hospitals were, for instance, possible candidates for public anatomies.

26 Park, "Criminal and Saintly Body," 14–15.

27 Lassek, *Human Dissection,* 32.

28 Sawday, *Body Emblazoned,* 63. It is important to make a distinction between northern and southern Europe when discussing attitudes towards anatomy. Sawday's book may be proposing a valid argument for northern Europe (although one may also think that Sawday sometimes stretches the evidence), but it is not quite as valid for southern Europe and Italy particularly. See, for instance, John Gouws's review of *The Body Emblazoned* in *Notes and Queries* 44 (1997): 556. Sawday's statement that the link between anatomy and public executions made anatomy a suspect profession approached with fascination and horror is particularly troublesome (even for northern Europe, but particularly for southern Europe). In fact, in Italy dissected bodies of criminals were often given a proper burial, especially if they repented before the execution. It is more a projection of our own contemporary view of anatomy and dissection than a historically accurate rendering of Renaissance perspectives. On attitudes towards the body and the soul in relation to dissection, see Katharine Park, "The Life of the Corpse: Division and Dissection in Late Medieval Europe," *Journal of the History of Medicine and the Allied Sciences* 50 (1995): 111–32.

29 As Katherine Park suggests in "The Criminal and the Saintly Body": "There are clear indications that anatomists sometimes eliminated the middleman by carrying out capital sentences themselves" (20).

30 Ruth Richardson, *Death, Dissection and the Destitute* (London: Routledge, 1987), 30– 51.

31 Practices varied from country to country. The integration of capital punishment and public dissection was not commonplace in all European countries. Italy was most reticent in this respect, England most explicit. For more details, see Carlino, *Books of the Body,* 219, and Samual Y. Edgerton, *Pictures and Punishment: Art and Criminal Prosecution during the Florentine Renaissance* (Ithaca, N.Y.: Cornell University Press, 1985).

32 This was not the case in all European cities; the anatomical theater De Waag, in Amsterdam, for instance, publicly dissected local criminals who were identified explicitly by listing their names and crimes. Britain and Italy had strict rules concerning the anonymity of the dissected corpse.

33 The cooperation between judges and anatomists is extensively described by Giovanni Ferrari, "Public Anatomy Lessons and the Carnival: The Anatomy Theatre of Bologna," *Past and Present* 117 (1987): 50–106.

34 See Andrea Carlino, *Books of the Body:* "The Body had to be someone who had been condemned to death . . . preferably a youthful body, in good condition, and of strong musculature, such as to permit a successful demonstration; the moral quality of the body had to be evaluated at the same time. Criminals had to be found guilty in a criminal court. Hanging was preferred, so the body was not disfigured by torture, punishment, mutilation or execution" (92).

35 Park, "Criminal and Saintly Body,"13

36 For a more elaborate description of the different uses of female and male corpses, see Ludmilla Jordanova, *Sexual Visions: Images of Gender in Science and Medicine between the Eighteenth and Twentieth Centuries* (Madison: University of Wisconsin Press, 1989). See also Katharine Park, "Dissecting the Female Body:

From Women's Secrets to the Secrets of Nature," in *Crossing Boundaries: Attending to Early Modern Women,* ed. Jane Donawerth and Adele Seeff (Newark: University of Delaware Press; London and Toronto: Associated University Presses, 2000), 29–45.

37 Medical and other science journals published extensive correspondence between specialists on the ethics of using the cadaver of a convicted criminal for virtual anatomy. See, for instance, Meredith Wadman, "Ethics Worries over Execution Twist to Internet's 'Visible Man,'" *Nature* 382 (1996): 657; Mitchell M. Waldrop, "The Visible Man Steps Out," *Science* 269 (1995): 1358; and Stuart Martin, "Concentrating the Mind," *Nature* 383 (1996): 381.

38 An interesting analysis of newspaper reactions to the Visible Male was given by Thomas Csordas, "Computerized Cadavers: Shades of Being and Representation in Virtual Reality" (paper presented at the conference Biotechnology, Culture, and the Body, University of Wisconsin, Milawaukee, April 1997).

39 Thomas Csordas, "Computerized Cadavers," cites headlines of newspapers lamenting the reanimation of Jernigan, such as "Executed Killer Reborn as Visible Man" and "Killer Let Loose on the Internet."

40 Besides the Visible Male and Visible Female, scientists are creating a Visible Embryo, a project supervised by the Armed Forces Institute of Pathology; the database of the Visible Embryo will use as its material basis a collection of embryos from the Carnegie Mellon Collection of Human Embryology, which was a gift to the National Museum of Health and Medicine in Washington, D.C., by the German embryologist Erich Blechschmidt of the University of Goettingen, Germany. See Julie A. Miller, "Anatomy via the Internet: Visible Human Project and Visible Embryo Project, *BioScience* 44 (1994): 397; and Philip Cohen, "Tough Gestation for Virtual Embryo," *New Scientist* 152 (1996): 6.

41 The Visible Male database comprises of 15 gigabytes, and the Visible Female, 39 gigabytes.

42 For an insightful analysis of the gender-specificity of the Visible Female, see Lisa Cartwright, "A Cultural Anatomy of the Visible Human Project," in *The Visible Woman: Imaging Technologies, Gender, and Science,* eds. Paula A. Treichler, Lisa Cartwright, and Constance Penley (New York: New York University Press, 1998), 21–43.

43 Bologna was the first town to have an anatomical theater, built in 1595; other cities (Padova, Leiden, and Amsterdam) followed in the seventeenth century. See Giovanni Ferrari, "Public Anatomy Lessons and the Carnival," 72.

44 On the frontispiece of Vesalius's most famous work, *De Humani Corporis Fabrica* (1543), we can see how a large and varied crowd observes his dissection. For a detailed analysis of this frontispice, see Carlino, *Books of the Body,* chapter 2.

45 See, for instance, Giovanni Ferrari, "Public Anatomy Lessons and the Carnival," pp. 68–78.

46 This detail is provided by Luke Wilson, "William Harvey's 'Prelectiones,'" 68–69.

47 A year after their release on the Internet, the Visible Human databases were being used by some 500 institutions around the world, and the number of users is still growing. This success was reported

by Denise Grady, "Research Uses Grow for Virtual Cadavers," *New York Times,* October 8, 1996.

48 The VOXEL-MAN Project has transformed the Visible Human databases into user-friendly products, such as CD-ROMS and videos, which can be used at various educational levels. The CD-ROM VOXEL-MAN Junior: Interactive Anatomy and Radiology in Virtual Reality Scenes (New York: Springer Verlag, 1998), for instance, is a reconstruction of the Visible Male's head; the CD-ROM is suitable for (high-school) students, but is of little use to medical professionals.

49 *Body Vovage* (New York: Time Warner Electronic Publishing, 1997), software by Learn Technologies, interactive content by Alexander Tsiaras. The book accompanying this CD-ROM contains beautiful full-color scans, and is published by the same company.

50 Don Ihde, "Virtual Bodies," in *Body and Flesh: A Philosophical Reader,* ed. Don Welton (London: Blackwell, 1998), 349–57, raises the question as to whether "virtual reality" could eventually replace "real life" in the many fields where virtual reality is currently being tested. He outlines the long and protracted history of this question, and ends on a revealing note: "Could VR replace RL? Only if theater could replace actual life!" (355).

51 Simon J. Williams, "Modern Medicine and the Uncertain Body: From Corporeality to Hyperreality," *Social Science and Medicine* 45 (1997): 1041–49, raises the general question as to whether hyperreality may be replacing corporeality in the future, and explains digital dissection in the larger context of medical technologies, such as new reproductive technologies and telemedicine.

EPILOGUE

1 For a brief introduction into the visual ethics of photography, television, and film, see Larry Gross, John S. Katz, and Jay Ruby, eds. *Image Ethics: The Moral Rights of Subjects in Photographs, Film, and Television* (Oxford: Oxford University Press, 1988).

2 See especially Brian Winston, "The Tradition of the Victim in Griersonian Documentary," in Gross, Katz, and Ruby, *Image Ethics,* 34–57; see also Caroline Anderson and Thomas Benson, "Direct Cinema and the Myth of Informed Consent: The Case of Titicut Follies," in ibid., 58–90.

3 In this context, Barbara Stafford, *Good Looking: Essays on the Virtue of Images* (Cambridge, Mass.: MIT Press, 1996), refers to the "urgent need to develop means for educating diverse audiences to the true, false, and ambiguous dimensions of pictures, high and low" (87).

4 For an excellent analysis of the interrelationship between medical-scientific culture and visual culture in general, see Marita Sturken and Lisa Cartwright, *Practices of Looking: An Introduction to Visual Culture* (Oxford: Oxford University Press, 2001), chapter 8.

BIBLIOGRAPHY

Ackerman, Michael J. "The Visible Human Project: A Resource for Education." *Academic Medicine* 74 (1999): 667–70.

Adler, Kathleen, and Marcia Pointon, ed. *The Body Imaged: The Human Form and Visual Culture since the Renaissance.* Cambridge: Cambridge University Press, 1993.

Akrich, Madeleine, and Bernike Pasveer. "Passages of Surveillance and Coordination." *Theoretical Medicine* 21 (2000): 63–83.

Amann, K., and K. Knorr Cetina. "The Fixation of (Visual) Evidence." In *Representation in Scientific Practice,* edited by Michael Lynch and Steve Woolgar, 85–122. Cambridge, Mass.: MIT Press, 1988.

Anderson, Caroline, and Thomas Benson. "Direct Cinema and the Myth of Informed Consent: The Case of Titicut Follies." In *Image Ethics: The Moral Rights of Subjects in Photographs, Film, and Television,* edited by Larry Gross, John S. Katz, and Jay Ruby, 58–90. Oxford: Oxford University Press, 1988.

Arnold, Jean-Michel. "La Grammaire Cinématographique: Une Invention des Scientifiques." In *Le Cinéma et la Science,* edited by A. Martinet, 211–17. Paris: CNRS, 1994.

Ayers, S., and A. D. Pickering. "Psychological Factors and Ultrasound: Differences between Routine and High-risk Scans." *Ultrasound in Obstetrics and Gynaecology* 9 (1997): 76–79.

Baillie, C. "Should Utrasound Scanning in Pregnancy Be Routine?" *Journal of Reproductive and Infant Psychology* 17 (1999): 149–57.

Baillie, C., G. Mason, and J. Hewison. "Scanning for Pleasure." *British Journal of Obstetrics and Gynaecology* 104 (1997): 1223–24.

Bal, Mieke, ed. *The Practice of Cultural Analysis: Exposing Interdisciplinary Interpretation.* Stanford: Stanford University Press, 1999.

Balsamo, Anne. *Technologies of the Gendered Body: Reading Cyborg Women.* Durham: Duke University Press, 1996.

Barad, Karen. "Getting Real: Technoscientific Practices and the Materialization of Reality." *Differences* 10 (1998): 87–128.

Bates, Barbara. *Bargaining for Life: A Social History of Tuberculosis, 1876–1938.* Philadelphia: University of Pennsylvania Press, 1992.

Beaulieu, Anne. "The Brain at the End of the Rainbow: The Promises of Brain Scans in the Research Field and in the Media." In *Wild Science: Reading Feminism. Medicine and the Media,* edited by Janine Marchessault and Kim Sawchuk, 39–54. New York: Routledge, 2000.

Belling, Catherine. "Reading *The Operation:* Television, Realism, and the Possession of Medical Knowledge." *Literature and Medicine* 17 (1998): 1–23.

Berg, Marc, and Annemarie Mol, eds. *Differences in Medicine: Unraveling Practices, Techniques, and Bodies.* Durham: Duke University Press, 1998.

Berkowitz, R. W. "Should Every Pregnant Woman Undergo Ultrasonography?" *New England Journal of Medicine* 329 (1993): 874–75.

Bijker, Wiebe, and John Law. *Shaping Technology/Building Society.* Cambridge, Mass.: MIT Press, 1992.

Black, Rita B. "Seeing the Baby: The Impact of Ultrasound Technology." *Journal of Genetic Counseling* 1 (1992): 45–54.

Blume, Stuart. *Insight and Industry: On the Dynamics of Technological Change in Medicine.* Cambridge, Mass.: MIT Press, 1992.

Bogdan, Robert. *Freak Show: Presenting Human Oddities for Amusement and Profit.* Chicago: University of Chicago Press, 1988.

Bolter, Jay D., and Richard Grusin. *Remediation: Understanding New Media.* Cambridge, Mass.: MIT Press, 1998.

Bond, John H. "Virtual Colonoscopy—Promising but Not Ready for Widespread Use." *New England Journal of Medicine* 341 (November 11, 1999): 1540–42.

Braun, Marta. *Picturing Time: The Work of Etienne-Jules Marey, 1830–1904.* Chicago: Chicago University Press, 1992.

Bryder, Linda. *Below the Magic Mountain: A Social History of Tuberculosis in Twentieth-Century Britain.* Oxford: Clarendon Press, 1988.

Campbell, S., and E. Reading. "Ultrasound Scanning in Pregnancy: The Short-term Psychological Effects of Early Real-time Scans." *Journal of Psychosomatic Obstetrics and Gynaecology* 1 (1982): 57–60.

Carlino, Andrea. *Books of the Body: Anatomical Ritual and Renaissance Learning.* Chicago: Chicago University Press, 1999.

Cartwright, Lisa. "A Cultural Anatomy of the Visible Human Project." In *The Visible Woman: Imaging Technologies, Gender, and Science,* edited by Paula A. Treichler, Lisa Cartwright, and Constance Penley, 21–43. New York: New York University Press, 1998.

———. *Screening the Body: Tracing Medicine's Visual Culture.* Minneapolis: University of Minnesota Press, 1995.

Chervenak, Frank A., and Laurence B. McCullough. "Ethical Dimensions of Ultrasound Screening for Fetal Anomalies." In *Ultrasound Screening for Fetal Anomalies: Is It Worth It?* edited by Salvatore Levi and Frank Chervenak, 185–90. New York: New York Academy of Sciences, 1998.

Chudleigh, Trish. "Scanning for Pleasure." *Ultrasound in Obstetrics and Gynaecology* 14 (1999): 369–71.

Clark, David, and Catherine Myser. "Being Humaned: Medical Documentaries and the Hyperrealization of Conjoined Twins." In *Freakery: Cultural Spectacles of the Extraordinary Body,* edited by Rosemarie G. Thomson, 338–55. New York: New York University Press, 1994.

Cohen, Philip. "Tough Gestation for Virtual Embryo." *New Scientist* 152 (1996): 6.

Connor, Steve. "Integuments: The Scar, the Sheen, the Screen." *New Formations* 39 (1999): 32–54.

Craig, Marveen. "Controversies in Obstetric Gynecologic Ultrasound." In *Diagnostic Medical Sonography: A Guide to Clinical Practice,* edited by M. Berman. Philadelphia: Lippincott, 1991.

Crary, Jonathan. *Techniques of the Observer: On Vision and Modernity in the Nineteenth Century.* Cambridge, Mass.: MIT Press, 1990.

Crawford, T. Hugh. "Imaging the Human Body: Quasi Objects, Quasi Texts, and the Theater of Proof." *PMLA* 111 (1996): 66–79.

———. "Visual Knowledge in Medicine and Popular Film." *Literature and Medicine* 17 (1998): 24–44.

Cushieri, A., and O. Berci. *Laparoscopic Biliary Surgery.* Oxford: Blackwell, 1990.

Cussins, Charis M. "Ontological Choreography: Agency for Women Patients in an Infertility Clinic." In *Differences in Medicine,* edited by Marc Berg and Annemarie Mol, 166–201. Durham: Duke University Press, 1998.

Dale, Karen. "Identity in a Culture of Dissection: Body, Self and Knowledge." In *Ideas of Difference: Social Spaces and the Labour of Division,* edited by Kevin Hetherington and Rolland Munro, 94–113. London: Blackwell, 1997.

Daston, Lorraine. "Objectivity and the Escape from Perspective." *Social Studies of Science* 22 (1992): 597–618.

Daston, Lorraine, and Peter Galison. "The Image of Objectivity." *Representations* 40 (1992): 81–128.

Daston, Lorraine, and Katherine Park. *Wonders and the Order of Nature, 1150–1170.* New York: Zone Press, 1998.

Debord, Guy. *The Society of the Spectacle.* Detroit: Black and Red, 1977.

Dermer, Rachelle A. "Joel-Peter Witkin and Dr. Stanley B. Burns: A Language of Body Parts." *History of Photography* 23 (1999): 245–53.

Dijck, José van. *ImagEnation: Popular Images of Genetics.* New York: New York University Press, 1998.

———. *Manufacturing Babies and Public Consent: Debating the New Reproductive Technologies.* New York: New York University Press, 1995.

Doby, T., and G. Alker. *Origins and Development of Medical Imaging.* Carbondale, Southern Illinois Press, 1997.

Doyen, Eugene-Louis. "Le Cas des Xiphopages Hindoues Radica-Doodica." *Revue Critique de Médicine et de Chirurgie* 4 (1902): 3031.

Dumit, Joseph. "Brain-Mind Machines and American Technological Dream Marketing." In *Cyborgs and Citadels: Anthropological Interventions in Emerging Sciences and Technology,* edited by Gary L. Downey and Joseph Dumit,

347–62. Santa Fe: School of American Research Press, 1999.

———. "Objective Brains, Prejudicial Images." *Science in Context* 12 (1999): 173–201.

———. "When Explanations Rest: 'Good-enough' Brain Science and the New Sociomedical Disorders." In *Living and Working with the New Medical Technologies: Intersections of Inquiry,* edited by Margaret Lock, Allan Young, and Alberto Cambriosio, 209–32. Cambridge: Cambridge University Press, 2000.

Edelmann, Claude. "A la Découverte du Corps Humain." In *Le Cinéma et la Science,* edited by A. Martinet, 174–81. Paris: CNRS, 1994.

Edgerton, Samuel Y. *Pictures and Punishment: Art and Criminal Prosecution during the Florentine Renaissance.* Ithaca, N.Y.: Cornell University Press, 1985.

Essex-Lopresti, Michael. "Centenary of the Medical Film." *The Lancet* 349 (1997): 819–20.

———. "The Medical Film, 1897–1997: Part I." *Journal of Audiovisual Media in Medicine* 21 (1998): 48–55.

Ewigman, Bernard G. "Effect of Prenatal Ultrasound Screening on Perinatal Outcome." *New England Journal of Medicine* 329 (1993): 821–29.

Ewigman, Bernard G., and J. P. Crane. "A Randomized Trial of Prenatal Ultrasonographic Screening: Impact on Maternal Management and Outcome." *American Obstetrics and Gynaecology* 169 (1993): 483–89.

Fenlon, Helen M. "A Comparison between Virtual and Conventional Colonoscopy for the Detection of Colorectal Polyps." *New England Journal of Medicine* 341 (November 11, 1999): 1496–503.

Ferrari, Giovanni. "Public Anatomy Lessons and the Carnival: The Anatomy Theatre of Bologna." *Past and Present* 117 (1987): 50–106.

Fiedler, Leslie. *Freaks: Myths and Images of the Secret Self.* New York: Simon and Schuster, 1978.

Fletcher, J. C., and M. I. Evans. "Maternal Bonding in Early Fetal Ultrasound Examinations." *New England Journal of Medicine* 308 (1983): 392–93.

Foucault, Michel. *The Birth of the Clinic: An Archeaology of Medical Perception.* London: Travistock, 1973.

Franklin, Sarah. "Postmodern Procreation: Representing Reproductive Practice." *Science as Culture* 3 (1993): 522–61.

French, Roger. *Dissection and Vivisection in the European Renaissance.* Aldershot: Ashgate, 1999.

Froehlich, F., and J. J. Gonvers. "Gastrointestinal Endoscopy: Do We Perform Too Many or Not Enough Procedures?" *Canadian Journal of Gastroenterology* 13 (1999): 345–46.

Gabbe, S. G. "Routine versus Indicated Scans." In *Diagnostic Ultrasound Applied to Obstetrics and Gynaecology,* edited by Ruddy E. Sabbagha. Philadelphia: Lippincott, 1994.

Garner, P., and M. Collins. "The Application of Telepresence in Medicine." In *Telepresence,* edited by P. J. Sheppard and G. R. Walker, 323–33. Boston: Kluwer, 1999.

Gerber, David. "The 'Careers' of People Exhibited in Freak Shows: The Problem of Volition and Valorization." In

Freakery: Cultural Spectacles of the Extraordinary Body, edited by Rosemarie G. Thomson, 38–54. New York: New York University Press, 1994.

Gerstenberger, P. D., and P. A. Plumeri. "Malpractice Claims in Gastrointestinal Endoscopy: Analysis of an Insurance Industry Data-base." *Gastrointestinal Endoscopy* 39 (1993): 132–38.

Gonzalez-Crussi, F. *Suspended Animation: Six Essays on the Preservation of Bodily Parts.* San Diego: Harcourt Brace, 1995.

Gorman, P. J., and A. H. Meier. "Simulation and Virtual Reality in Surgical Education—Real or Unreal?" *Archives of Surgery* 134 (November 1999): 1203–8.

Green, Josephine, and Helen Statham. "Psychosocial Aspects of Prenatal Screening and Diagnosis." In *The Troubled Helix: Social and Psychological Implications of the New Human Genetics,* edited by Theresa Marteau and Martin Richards, 140–58. Cambridge: Cambridge University Press, 1996.

Grimson, E. L. "Image-Guided Surgery." *Scientific American* (June 1999): 62–69.

John Gouws. Review of *The Body Emblazoned. Notes and Queries* 44 (1997): 556.

Gross, Larry, John S. Katz, and Jay Ruby, ed. *Image Ethics: The Moral Rights of Subjects in Photographs, Film, and Television.* Oxford: Oxford University Press, 1988.

Grosz, Elizabeth. "Intolerable Ambiguity: Freaks as/at the Limit." In *Freakery: Cultural Spectacles of the Extraordinary Body,* edited by Rosemarie G. Thomson, 55–66. New York: New York University Press, 1994.

Grosz, Elizabeth. *Volatile Bodies: Toward a Corporeal Feminism.* New York: Routledge, 1994.

Grove, Alan. "Röntgen's Ghosts: Photography, X-Rays, and the Victorian Imagination." *Literature and Medicine* 17 (1998): 141–73.

Gunning, Tom. "The Cinema of Attraction: Early Film, Its Spectator, and the Avant-garde." *Wide Angle* 8 (1986): 63–70.

Hacking, Ian. *Representing and Intervening: Introductory Topics in the Philosophy of Natural Science.* Cambridge: Cambridge University Press, 1995.

Hankins, Thomas L., and Robert J. Silverman. *Instruments and the Imagination.* Princeton: Princeton University Press, 1995.

Hansen, Julie V. "Resurrecting Death: Anatomical Art in the Cabinet of Dr. Frederick Ruysch." *Art Bulletin* 78 (1996): 663–79.

Haraway, Donna. *Simians, Cyborgs, and Women: The Reinvention of Nature.* New York, Routledge, 1991.

Harcourt, Glenn. "Andreas Vesalius and the Anatomy of Antique Sculpture." *Representations* 17 (1987): 28–61.

Hawkins, Joan. "One of Us: Tod Browning's *Freaks.*" In *Freakery: Cultural Spectacles of the Extraordinary Body,* edited by Rosemarie G. Thomson, 265–76. New York: New York University Press, 1994.

Hayles, N. Katherine. *How We Became Posthuman: Virtual Bodies in Cybernetics, Literature, and Informatics.* Chicago: University of Chicago Press, 1999.

Herzig, Rebecca. "Removing Roots: North American Hiroshima Maidens and the X-ray." *Technology and Culture* 40 (1999): 723–45.

Hillowala, Rummy. *The Anatomical Waxes of La Specola.* Florence: Arnaud, 1995.

Hirschauer, Stefan. "The Manufacture of Bodies in Surgery." *Social Studies of Science* 21 (1991): 279–319.

Hirschowitz, B. I. "Development and Application of Endoscopy." *Gastroenterology* 104 (1993): 337–42.

Holtzmann-Kevles, Bettyann. *Naked to the Bone: Medical Imaging in the Twentieth Century.* New Brunswick, N.J.: Rutgers University Press, 1997.

Huhtamo, Erkki. "Encapsulated Bodies in Motion." In *Critical Issues in Electronic Media,* edited by Simon Perry, 159–86. New York: SUNY Press, 1995.

Hunfeld, J. A. M. "Emotional Reactions in Late Pregnancy Following the Ultrasound Diagnosis of Severe or Lethal Fetal Malformation." *Prenatal Diagnosis* 13 (1993): 603–12.

Hutcheon, Linda. *The Politics of Postmodernism.* London: Routledge, 1989.

Hyde, Beverly. "An Interview Study of Pregnant Women's Attitudes to Ultrasound Screening." *Social Science and Medicine* 22 (1986): 587–92.

Ihde, Don. *Expanding Hermeneutics: Visualization in Science.* Chicago: Northwestern University Press, 1999.

———. "Virtual Bodies." In *Body and Flesh: A Philosophical Reader,* edited by Don Welton, 349–57. London: Blackwell, 1998.

Jain, Sarah. "Mysterious Delicacies and Ambiguous Agents: Lennart Nilsson in National Geographic." *Configurations* 6 (1998): 373–95.

Johnson, Martin P., and John E. Pudifoot. "Miscarriage: Is Vividness of Visual Imagery a Factor in the Grief Reaction of the Partner?" *British Journal of Health Psychology* 3 (1998): 137–46.

Jordanova, Ludmilla. "Medicine and the Genres of Display." In *Visual Display: Culture Beyond Appearances,* edited by Lynne Cooke and Peter Wollen, 202–17. Seattle: Bay Press, 1995.

———. *Sexual Visions: Images of Gender in Science and Medicine between the Eighteenth and Twentieth Centuries.* Madison: University of Wisconsin Press, 1989.

Kieler, Helle. *Effects and Possible Side Effects of Routine Ultrasound Scanning in Pregnancy.* Uppsala: Acta Universitatis Upsaliensis, 1997.

King, John S. "Most Dubious: Myth, the Occult, and Politics in the Zauberberg." *Monatshefte* 88 (1996): 217–36.

Kittler, Friedrich A. *Discourse Networks, 1800–1900.* Translated by Michael Metter. Stanford: Stanford University Press, 1990. Originally published as *Aufschreibesysteme, 1800–1900* (München: Wilhelm Fink Verlag, 1985).

———. *Gramophone, Film, Typewriter.* Translated by Geoffrey Winthrop-Young and Michael Wutz. Stanford: Stanford University Press, 1999. Originally published as *Grammophon Film Typewriter* (Berlin, Brinkmann & Bose, 1986).

Klaus, M., and P. Jerauld. "Maternal Attractions: Importance of the First Postpartum Days." *New England Journal of Medicine* 286 (1972): 460–63.

Koc, Richard. "Magical Enactments: Reflections on 'Highly Questionable' Matters in *Der Zauberberg.*" *Germanic Review* 68 (1993): 107–17.

Koch, Ellen. "In the Image of Science?

Negotiating the Development of Diagnostic Ultrasound in the Cultures of Surgery and Radiology." *Technology and Culture* 34 (1993): 858–93.

Lammer, Christina. *Die Puppe: Eine Anatomie des Blicks.* Vienna: Verlag Turia, 1999.

Lassek, A. M. *Human Dissection: Its Drama and Struggle.* Springfield, Ill.: Thomas, 1958.

Latour, Bruno. "Visualization and Cognition: Thinking with Eyes and Hands." In *Knowledge and Society: Studies of the Sociology of Culture,* edited by Henrika Kuklich and Elizabeth Lang, 1–40. London: Jai Press, 1986.

Lauridsen, Laurits. *Lanterna Magica in Corpore Humano.* Arhus, Denmark: Steno Museum, 1998.

Lefèbvre, Thierry. "La Collection des films du Dr. Doyen." *1895* 17 (1994): 100–114.

———. "Le docteur Doyen, un précurseur." In *Le Cinéma et la Science,* edited by A. Martinet, 70–77. Paris: CNRS, 1994.

Lenoir, Tim, and Xin Wei Sha. "Authorship and Surgery: The Shifting Ontology of the Virtual Surgeon." In *From Energy to Information,* edited by Brace Clark and Linda Henderson. Stanford: Stanford University Press, 2002.

Lerner, Barron H. "The Perils of X-Ray Vision: How Radiographic Images Have Historically Influenced Perception." *Perspectives in Biology and Medicine* 35 (1992): 382–97.

Levi, Salvatore. "Routine Ultrasound Screening of Congenital Anomalies: An Overview of the European Experience." In *Ultrasound Screening for Fetal Anomalies: Is It Worth It?* edited by Salvatore Levi and Frank Chervenak, 86–97. New York: New York Academy of Sciences, 1998.

Levi, Salvatore, and Frank Chervenak, eds. *Ultrasound Screening for Fetal Anomalies: Is It Worth It?* New York: New York Academy of Sciences, 1998.

Lévy, Pierre. *Becoming Virtual: Reality in the Digital Age.* New York: Plenum Trade, 1998.

Lindfors, Ben. "Ethnological Show Business: Footlighting the Dark Continent." In *Freakery: Cultural Spectacles of the Extraordinary Body,* edited by Rosemarie G. Thomson, 207–18. New York: New York University Press, 1994.

Lupton, Deborah. *Medicine as Culture: Illness, Disease and the Body in Western Societies.* London: Sage, 1994.

MacKenzie, C., and N. Stoljar. *Relational Autonomy.* Oxford: Oxford University Press, 2000.

Mann, Thomas. "Drei Berichte über Okkultische Sitzungen." *Gesammelte Werke.* Vol. 13: 33–48. Berlin: Fisher Verlag, 1965.

———. *The Magic Mountain.* Translated by Helen Tracy Lowe-Porter. New York: Random House, 1969. Originally published as *Der Zauberberg* (Fisher Verlag, 1924).

Marchessault, Janine, and Kim Sawchuk. *Wild Science: Feminist Images of Medicine and Body.* New York: Routledge, 2000.

Marescaux, J., and D. Mutter. "The Virtual University Applied to Telesurgery: From Tele-education to Tele-manipulation." *Bulletin de L'Académie Nationale de Médecine* 183 (1999): 509–22.

Martin, Emily. *Flexible Bodies: The Role of Immunity in American Culture from the Days of Polio to the Age of AIDS.* Boston: Beacon Press, 1994.

Martin, Stuart. "Concentrating the Mind." *Nature* 383 (1996): 381.

McFayden, Anne. "First Trimester Ultrasound Screening Carries Ethical and Psychological Implications." *British Medical Journal* 317 (1998): 694–95.

McNay, Margareth B., and John E. Fleming. "Forty Years of Obstetric Ultrasound, 1957–97: From A-scope to Three Dimensions." *Ultrasound in Medicine and Biology* 25 (1999): 3–56.

Miller, Julie A. "Anatomy via the Internet: Visible Human Project and Visible Embryo Project." *BioScience* 44 (1994): 397.

Mitchell, William J. *The Reconfigured Eye: Visual Truth in the Photographic Era.* Cambridge, Mass.: MIT Press, 1992.

Montgomery, Scott. *The Scientific Voice,* New York: The Guilford Press, 1996.

Morrin, Martina M. "Virtual Colonoscopy: A Kinder, Gentler Colorectal Cancer Screening Test?" *The Lancet* 354 (September 25 1999): 1048–49.

Mulvey, Laura. "Visual Pleasures and Narrative Cinema." *Screen* 16 (1975): 6–18.

Myser, Catherine, and David Clark. "Fixing Katie and Eilish: Medical Documentaries and the Subjection of Conjoined Twins." *Literature and Medicine* 17 (1998): 45–67.

Nardie, Bonnie. "Video-as-Data: Technical and Social Aspects of a Collaborative Multimedia Approach." In *Video-Mediated Communication,* edited by Abigail Sallen, Sylvia Wilbur, and Kathleen Fink, 487–517. Mahwah, N. J.: Lawrence Elbaum, 1997.

Natali, J. "Medicolegal Implications of Vascular Injuries during Videoendoscopic Surgery." *Journal Des Maladies Vasculaires* 21 (1996): 223–26.

Newman, Karen. *Fetal Positions: Individualism, Science, Visuality.* Stanford: Stanford University Press, 1996.

Newman, Paul G., and Grace S. Rozycki. "The History of Ultrasound." *Surgical Clinics of North America* 78 (1998): 179–95.

Nutton, Vivian. "Representation and Memory in Renaissance Anatomical Illustration." In *Immagini per conoscere: dal Rinascimento alla Rivoluzione scientifica,* edited by Fabrizio Meroi and Claudio Pogliano, 61–80. Florence: L. S. Olschki Editore, 2001.

O'Connor, Erin. "Camera Medica: Towards a Morbid History of Photography." *History of Photography* 23 (1999): 232–44.

Ostman, Ronald E. "Photography and Persuasion: Farm Security Administration Photographs of Circus and Carnival Sideshows, 1935–42." In *Freakery: Cultural Spectacles of the Extraordinary Body,* edited by Rosemarie G. Thomson, 121–36. New York: New York University, 1994.

Oudshoorn, Nelly. *Beyond the Natural Body: An Archaeology of Sex Hormones.* London: Routledge, 1994.

Panofsky, Erwin. "Artist, Scientist, Genius: Notes on the 'Renaissance-Dämmerung.'" In *The Renaissance: Six Essays,* edited by Wallace K. Ferguson, 123–82. New York: Academy Library, 1953.

Park, Katherine. "The Criminal and the Saintly Body: Autopsy and Dissection in Renaissance Italy." *Renaissance Quarterly* 1 (1994): 1–33.

———. "Dissecting the Female Body: From Women's Secrets to the Secrets of Nature." In *Crossing Boundaries: Attending to Early Modern Women,* edited by Jane Donawerth and Adele Seeff, 29–45. Newark: University of Delaware Press; London and Toronto: Associated University Presses, 2000.

———. "The Life of the Corpse: Division and Dissection in Late Medieval Europe." *Journal of the History of Medicine and the Allied Sciences* 50 (1995): 111–32.

Park, Katherine, and Lorraine Daston. "Unnatural Conceptions: The Study of Monsters in Sixteenth- and Seventeenth Century France and England." *Past and Present* 92 (1981): 23–46.

Pasveer, Bernike. "Knowledge of Shadows: The Introduction of X-Ray Images in Medicine." *Sociology of Health and Illness* 11 (1989): 360–81.

———. "Wiens lijf? Wiens Leven? Echografie en het lichaam." In *De Nieuwe Mens: De Maakbaarheid van Lijf en Leven,* edited by F. de Lange, 43–57. Nijmegen: Thomas Moore Academie, 2000.

Petchesky, Rosalind. *Abortion and Woman's Choice: The State, Sexuality, and Reproductive Freedom.* Boston: Northeastern University Press, 1984.

Pingree, Allison. "America's 'United Siamese Brothers': Chang and Eng and Nineteenth-Century Ideologies of Democracy and Domesticity." In *Monster Theory: Reading Culture,* edited by Jeffrey J. Cohen. Minneapolis: University of Minnesota Press, 1996.

Ploeg, Irma van der. *Prosthetic Bodies: Female Embodiment in Reproductive Technologies.* Maastricht: Maastricht University Press, 1998.

Porter, Roy. *The Greatest Benefit to Mankind: A Medical History of Humanity from Antiquity to the Present.* London: Harper Collins, 1997.

Price, Francis. "Now You See It, Now You Don't: Mediating Science and Managing Uncertainty in Reproductive Medicine." In *Misunderstanding Science? The Public Reconstruction of Science and Technology,* edited by Alan Irwin and Brian Wynne, 84–106. Cambridge: Cambridge University Press, 1996.

Proust, Marcel. *Swann's Way.* Translated by Scott Moncrieff. London: Chatto and Windus, 1943.

Raskin, Valerie D. "Influence of Ultrasound on Parent's Reaction to Perinatal Loss." *American Journal of Psychiatry* 146 (1989): 1646.

Reed, T. J. *Thomas Mann: The Uses of Tradition.* Oxford: Oxford University Press, 1974.

Reiser, Stanley J. *Medicine and the Reign of Technology,* Cambridge: Cambridge University Press, 1978.

———. "Technology and the Use of the Senses in Twentieth-Century Medicine." In *Medicine and the Five Senses,* edited by Catherine F. Byrum and Roy Porter, 262–73. Cambridge: Cambridge University Press, 1996.

Richardson, Ruth. *Death, Dissection and the Destitute.* London: Routledge, 1987.

Robb, Richard A., and S. Aharon.

"Patient-specific Anatomic Models from 3-dimensional Medical Image Data for Clinical Applications in Surgery and Endoscopy." *Journal of Digital Imaging* 10 (1997): 31–35.

Roberts, K. B. *Maps of the Body: Anatomical Illustration through Five Centuries,* St. John's Memorial University of Newfoundland Press, 1981.

Roberts, K. B., and J. D. W. Tomlinson. *The Fabric of the Body: European Traditions of Anatomical Illustration.* Oxford: Clarendon Press, 1992.

Robertson, George, Melinda Mash, Lisa Tickner, Jon Bird, Barry Curtis, and Tim Putnam, eds. *Future Natural: Nature, Science, Culture.* London: Routledge, 1996.

Romano, Patril, and Norman J. Waitzman. "Can Decision Analysis Help Us Decide Whether Ultrasound Screening for Fetal Abnormalities is Worth It?" In *Ultrasound Screening for Fetal Anomalies: Is It Worth It?* edited by Salvatore Levi and Frank Chervenak, 154–68. New York: New York Academy of Sciences, 1998.

Rothman, Sheila M. *Living in the Shadow of Death: Tuberculosis and the Social Experience of Illness in American History.* New York: Harper, 1994.

Rowe, Katherine. "God's Handy Worke." In *The Body in Parts: Fantasies of Corporeality in Early Modern Europe,* edited by David Hillman and Carla Mazzio, 285–312. London: Routledge, 1997.

Rowe, Paul M. "Visible Human Project Pays Back Investment." *The Lancet* 352 (1999): 46.

Saari-Kemprainen A., and O. Karjalainen. "Ultrasound Screening and Perinatal Mortality: Controlled Trial of Systematic One-Stage Screening in Pregnancy; The Helsinki Ultrasound Trial." *The Lancet* 336 (1990): 387–91.

Sandelowski, Margareth. "Separate, but Less Unequal: Fetal Ultrasonography and the Transformation of Expectant Mother/Fatherhood." *Gender and Society* 8 (1994): 230–45.

Satava, Richard M. *Cybersurgery: Advanced Technologies for Surgical Practice.* New York: Wiley-Liss, 1998.

———. "Medical Virtual Reality: The Current State of the Future." In *Health Care in the Information Age,* edited by S. J. Weghorst, H. B. Sieburg, and K. S. Morgan, 100–106. Amsterdam: IOS Press, 1996.

Sawchuk, Kim. "Biotourism, Fantastic Voyage, and Sublime Inner Space." In *Wild Science: Feminist Images of Medicine and Body,* edited by Janine Marchessault and Kim Sawchuk, 9–23. New York: Routledge, 2000.

Sawday, Jonathan. *The Body Emblazoned: Direction and the Human Body in Renaissance Culture.* London: Routledge, 1995.

Schnalke, Thomas. *Disease in Wax: The History of Medical Moulage.* London: Quintessence, 1995.

Schofield, Stanley, and Michael Essex-Lopresti. "Filming the Surgical Separation of the Conjoined Twins of Kano." *Science and Film* 3 (1954): 19–23.

Schwartz, Vanessa. "Cinematic Spectatorship before the Apparatus: The Public Taste for Reality in Fin-de-Siècle Paris." In *Viewing Positions: Ways of Seeing Film,* edited by Linda Williams, 87–113. New Brunswick, N.J.: Rutgers University Press, 1995.

Schwarz, Hillel. *The Culture of the Copy: Striking Likeness, Unreasonable Facsimiles.* New York: Zone Books, 1996.

Seematter-Bagnoud, L., and J. Vader. "Overuse and Underuse of Diagnostic Upper Gastrointestinal Endoscopy in Various Clinical Settings." *International Journal for Quality in Health Care* 11 (1999): 301–8.

Shohat, Ella. "Lasers for Ladies: Endo Discourse and the Inscription of Science." In *The Visible Woman: Imaging Technologies, Gender, and Science,* edited by Paula Treichler, Lisa Cartwright, and Constance Penley, 240–72. New York: New York University Press, 1998.

Siraisi, Nancy. "Anatomizing the Past: Physicians and History in Renaissance Culture." *Renaissance Quarterly* 53 (2000): 1–30.

———. "Vesalius and the Reading of Galen's *Teleology.*" *Renaissance Quarterly* 50 (1997): 1–37.

Skupsi, D. W., F. A. Chervenak, and M. McCullough. "Routine Obstetric Ultrasound." *International Journal of Gynaecology and Obstetrics* 50 (1995): 233–42.

Sontag, Susan. *Illness as Metaphor.* New York: Farrar, Straus and Giroux, 1977.

Stacey, Jackie. *Teratologies: A Cultural Analysis of Cancer.* London: Routledge, 1998.

Stafford, Barbara. *Body Criticism: Imaging the Unseen in Enlightenment Art and Medicine.* Cambridge, Mass.: MIT Press, 1993.

———. *Good Looking: Essays on the Virtue of Images.* Cambridge, Mass.: MIT Press, 1996.

Sturken, Marita, and Lisa Cartwright. *Practices of Looking: An Introduction to Visual Culture.* Oxford: Oxford University Press, 2001.

Taylor, Janelle C. "Image of Contradiction: Obstetrical Ultrasound in American Culture." In *Reproducing Reproduction: Kinship, Power and Technological Innovation,* edited by Sarah Franklin and Helena Ragone, 15–45. Philadelphia: University of Pennsylvania Press, 1997.

Teichmann, Yona, and Dorit Rabinovitz. "Emotional Reactions of Pregnant Women in Ultrasound Scanning and Postpartum." In *Stress and Anxiety,* edited by Charles Spielberger and Irwin G. Sarason. New York: Hemisphere Press, 1991.

Thacker, Eugene. "Performing the Technoscientific Body: Real Video Surgery and the Anatomy Theater." *Body and Society* 5 (1999): 317–36.

———."Visible Human: Digital Anatomy and the Hypertexted Body." *C-Theory* 21 (1998): no page numbers (web version).

———. "What Is Biomedia?" *Configurations* 11 (2003): 47–79.

Thomasma, David. "The Ethics of Caring for Conjoined Twins: The Lakeberg Twins." *Hastings Center Report* 26 (1992): 4–12.

Thompson, John B. *The Media and Modernity: A Social Theory of the Media.* Cambridge: Polity Press, 1995.

Thomson, Rosemarie G. *Freakery: Cultural Spectacles of the Extraordinary Body.* New York: New York University Press, 1994.

Trumbo, Jane. "Visual Literacy and Science Communication." *Science Communication* 20 (1999): 409–25.

Vaughan, Christopher A. "Ogling Igorots: The Politics and Commerce of Exhibiting Cultural Otherness." In *Freakery: Cultural Spectacles of the Extraordinary Body,* edited by Rosemarie G. Thomson, 219–33. New York: New York University Press, 1994.

Verny, Thomas R. "Obstetrical Procedures: A Critical Examination of Their Effect on Pregnant Women and Their Unborn and Newborn Children." *Pre- and Peri-Natal Psychology Journal* 7 (1992): 101–12.

Vogel, W. de, J. W. Varossieau, and R. Blumenthal-Rothschild. "The Film of the Frisian Conjoined Twins." *Science and Film* 3 (1954): 20–25.

Wadman, Meredith. "Ethics Worries over Execution Twist to Internet's 'Visible Man.'" *Nature* 382 (1996): 657.

Waldby, Catherine. "The Visible Human Project: Data into Flesh, Flesh into Data." In *Wild Science: Feminist Images of Medicine and Body,* edited by Janine Marchessault and Kim Sawchuk, 24–38. New York: Routledge, 2000.

Waldrop, Mitchell M. "The Visible Man Steps Out." *Science* 269 (1995): 1358.

Wallace, Irving, and Amy Wallace. *The Two.* New York: Simon and Schuster, 1978.

Warren, Frank. *Television in Medical Education Handbook.* Washington: American Medical Association, 1955.

Weir, Lorna. "Cultural Intertexts and Scientific Rationality: The Case of Pregnancy Ultrasound." *Economy and Society* 27 (1998): 249–58.

Wetz, Frans J., and Brigitte Tag, eds. *Schöne Neue Körperwelten: Der Streit um die Ausstellung.* Stuttgart: Klett-Cotta, 2001.

Williams, Simon J. "Modern Medicine and the Uncertain Body: From Corporeality to Hyperreality." *Social Science and Medicine* 45 (1997): 1041–49.

Wilson, Luke. "William Harvey's 'Prelectiones': The Performance of the Body in the Renaissance Theater of Anatomy." *Representations* 17 (1987): 62–95.

Winston, Brian. "The Tradition of the Victim in Griersonian Documentary." In *Image Ethics: The Moral Rights of Subjects in Photographs, Film, and Television,* edited by Larry Gross, John S. Katz, and Jay Ruby, 34–57. Oxford: Oxford University Press, 1988.

Wolbarst, Anthony B. *Looking Within: How X-Ray, C.T., M.R.I., Ultrasound, and Other Medical Images Are Created and How They Help Physicians Save Lives.* Berkeley: University of California Press, 1999.

Woods, Tim. *Beginning Postmodernism.* Manchester: Manchester University Press, 1999.

Woolgar, Steve, and Michael Lynch, eds. *Representation in Scientific Practice.* Boston: MIT Press, 1990.

Yoxen, Edward. "Seeing with Sound: A Study of the Development of Medical Images." In *The Social Construction of Technological Systems: New Directions in the Sociology and History of Technology,* edited by Wiebe Bijker, Thomas Hughes, and Trevor Pinch, 281–303. Cambridge, Mass.: MIT Press, 1987.

INDEX